Cambridge E

T0093325

Elements in the Philosophy of Biology
edited by
Grant Ramsey
KU Leuven
Michael Ruse
Florida State University

THE CHALLENGE OF EVOLUTION TO RELIGION

Johan De Smedt
Saint Louis University

Helen De Cruz
Saint Louis University

CAMBRIDGE
UNIVERSITY PRESS

CAMBRIDGE
UNIVERSITY PRESS

University Printing House, Cambridge CB2 8BS, United Kingdom

One Liberty Plaza, 20th Floor, New York, NY 10006, USA

477 Williamstown Road, Port Melbourne, VIC 3207, Australia

314–321, 3rd Floor, Plot 3, Splendor Forum, Jasola District Centre,
New Delhi – 110025, India

79 Anson Road, #06–04/06, Singapore 079906

Cambridge University Press is part of the University of Cambridge.

It furthers the University's mission by disseminating knowledge in the pursuit of
education, learning, and research at the highest international levels of excellence.

www.cambridge.org
Information on this title: www.cambridge.org/9781108716048
DOI: 10.1017/9781108685436

First published 2020

A catalogue record for this publication is available from the British Library.

ISBN 978-1-108-71604-8 Paperback
ISSN 2515-1126 (online)
ISSN 2515-1118 (print)

The Challenge of Evolution to Religion

Elements in the Philosophy of Biology

DOI:10.1017/9781108685436
First published online: January 2020

Johan De Smedt
Saint Louis University

Helen De Cruz
Saint Louis University

Author for correspondence: Johan De Smedt, johanvdesmedt@gmail.com

Abstract: We focus on three challenges of evolution to religion: teleology, human origins, and the evolution of religion itself. First, religious worldviews tend to presuppose a teleological understanding of the origins of living things, but scientists mostly understand evolution as nonteleological. Second, religious and scientific accounts of human origins do not align in a straightforward sense. Third, evolutionary explanations of religion, including religious beliefs and practices, may cast doubt on their justification. We show how these tensions arise and offer potential responses for religion. Individual religions can meet these challenges if some of their metaphysical assumptions are adapted or abandoned.

Keywords: evolution, teleology, human origins, cognitive science of religion

ISBNs: 9781108716048 (PB), 9781108685436 (OC)
ISSNs: 2515-1126 (online), 2515-1118 (print)

Contents

1 Science, Religion, and Evolution

1.1 An Asymmetric Tension

Religion and science are in a peculiar tension because they have an asymmetric dependence relationship. Some fundamental metaphysical presuppositions of science are religious in origin. At the same time, science questions these religious assumptions. To see this tension at work, consider miracles and their relationship to laws of nature. The Christian doctrine of creation holds that an intelligent creator designed the world according to intelligible and orderly laws. This conception of the world as governed by laws that are discoverable by human minds inspired natural philosophers in the early seventeenth century, such as René Descartes, Robert Boyle, and Isaac Newton, to formulate their mechanistic conceptions of the world. Some natural philosophers, such as Bernard Nieuwentyt and John Ray, believed that experimental science provides clear evidence for divine design: it shows how ingenious and intelligently designed the laws of nature are. An unintended consequence of this line of thinking, however, was that miracles – a central element of Christian doctrine, for instance, in the resurrection of Jesus of Nazareth – had become all but impossible. Miracles constituted, by definition, a violation of the laws of nature, although there were minority views such as Samuel Clarke's that saw miracles as merely surprising or unusual. The laws of nature admitted no exceptions. This gave rise to an unstable conception of miracles as events that violated immutable laws of nature (Harrison, 1995). As a result, miracles became improbable and testimony to such events highly suspect, as exemplified in David Hume's (1748) argument against miracles.

The Fall narrative provides another example of how science undermines the religious foundations it was built on. The Fall is a religious doctrine based on chapter 3 of the biblical book of Genesis, where the first man and woman disobey God by eating the fruit of the Tree of Knowledge. Through this act of rebellion, humans marred the image of God in them, and death entered the world. Early modern natural philosophers believed that our mental faculties and senses – which were allegedly superior before the Fall – degenerated as a result of this original sin. This idea of human depravity contributed to the rise of empirical science in the early modern period: if we cannot trust our reason or senses, we need to do experiments to find out more about the world, supplementing our limited sensory capacities with instruments such as telescopes and microscopes (Harrison, 2007). We need to study nature to find out what it looks like rather than start from a priori assumptions, which are unreliable due to our defective cognition. Empirical research, however, has cast serious doubt on the historicity of the Fall (see Section 3). The discovery of fossils of prehistoric animals indicated that death existed before humans did. Early geologists such as Nicolas Steno (1638–1686) pointed out that fossils could not be explained away as somehow formed within rocks: they genuinely were the remains of living organisms, many of which are now extinct (Murray, 2008: 79). In the nineteenth century, the evolution of humans from hominin ancestors and their continuity with other great apes challenged the idea of an original state of superior senses and intellect. Therefore, while the idea of the Fall may have encouraged empirical research, that research indicates there never was a historical Fall.

Scientific concepts and attitudes may have religious origins, but religious beliefs do not seem to play a direct, productive role in science today, and this difference in epistemic standing is reflected in the contemporary science and religion literature. A number of contemporary theologians engage earnestly with scientific theories (e.g., van Huyssteen, 2006). Scientists, however, rarely explicitly draw on religious concepts for their ideas. There are some exceptions; for example, the evolutionary biologist Theodosius Dobzhansky (1973: 127) drew on his Eastern Orthodox faith to make evolution intelligible to himself. Evolution through natural selection, he argued, made more sense as a method of creation than creating species through intelligent design only to let the vast majority of them die out: "The organic diversity becomes, however, reasonable and understandable if the Creator has created the living world not by caprice but by evolution propelled by natural selection."

This limited engagement between science and religion in the form of dialogue[1] and especially integration may in part be a historical accident. Scientists since the nineteenth century have increasingly adhered to methodological naturalism, as exemplified in initiatives such as the X Club. The X Club was founded in 1864 by Thomas Huxley and friends. Its aim was to promote science untrammeled by religious dogmas and also to professionalize science, making it less a hobby project for amateur clergymen and more a profession with full-time salaried scientists (Garwood, 2008). There might be a more fundamental tension at work, however, that is not merely a historical accident. Some authors (e.g., Dawes, 2016) hold that religious dogmatism inevitably conflicts with scientific fallibilism – a conflict that surpasses individual scientists and religious thinkers. As we will see in the following sections, conflict positions such as Dawes', which stress the epistemic inflexibility of religion, do not take sufficiently into account how far religious positions can shift to accommodate results of the sciences. Sophisticated models of divine action (Section 2) and of human evolution (Section 3) are sensitive to empirical

---

[1] We here follow Barbour's (2000) distinction between dialogue, integration, conflict, and independence as potential ways in which science and religion can relate.

evidence. When we consider evolutionary theory, it seems difficult to maintain a position of independence, which tries to restrict science to the domain of facts and religion to the domain of values (*pace* Gould, 2001).

Religion frequently does venture on the domain of facts, as illustrated by high-profile court cases on the teaching of creationism, notably the Scopes "Monkey" trial (1925) and the Kitzmiller versus Dover trial (2005). Creationism in the broad sense is the position that the world's structure can be adequately explained only by positing at least one creator god (Sedley, 2007). This position is not necessarily in contradiction to evolution, as some authors (e.g., Dobzhansky, 1973; Lamoureux, 2008) have argued that God creates by using the process of evolution. But some forms of "creationism" are in tension with evolution, and the term creationism is often used to denote creationist positions that deny that God uses evolution as a method of creation (Alexander, 2008).[2] In particular, the term usually refers to Old Earth creationism, which rejects macroevolution but accepts an old Earth (geology, cosmology), and to Young Earth creationism, which in addition to macroevolution also rejects findings of geology and cosmology by positing a literal six-day creation. Intelligent design creationism replaces the term "God" with a nonspecified intelligent designer but does not endorse macroevolution as a mode of creation either. Debates on whether teaching such forms of creationism in US public schools is constitutional will continue for the foreseeable future. Scientists, for their part, explore the domain of values. For example, the biochemist Jacques Monod (1970) argued that coming to terms with evolution meant grappling with a universe that was fundamentally amoral and indifferent to our hopes and fears. This indicates that a neat separation of scientific and religious views into views about facts and values (in the form of nonoverlapping magisteria) cannot be maintained.

1.2 Evolution and Religion: Examples from Judaism and Hinduism

Most of the philosophical and theological literature on evolutionary challenges to religion has focused on Christianity, and for this reason this Element will primarily draw on this tradition. Writers in other religious traditions, however, have also grappled with the challenge of evolution. To give a sense of these discussions, we here briefly look at Judaism and Hinduism.

Judaism has a long tradition of sophisticated engagement between science and religion, which informs debates on evolution to this day. For example, although he wrote centuries before the formulation of evolutionary theory, the

---

[2] Unless otherwise specified, the term creationism in this Element refers to the narrower definition of creationist views that exclude evolution as a method of creation.

theologian Maimonides (Rabbi Mōšeh ben-Maymūn, 1138–1204) has had an enduring influence on this topic. He aimed to reconcile Judaism with Aristotelian philosophy. His *Guide of the Perplexed* treats the potential sources of tension between the two and resists a wholesale acceptance of the Sages (religious experts responsible for formulating the Halakhah, or Jewish law) and a literalist interpretation of the Torah. Maimonides maintained that the Sages were experts on religious matters but that this expertise did not automatically extend to other domains. For example, he rejected astrology (which some Sages endorsed), arguing that it was not only scientifically dubious but also difficult to reconcile with human freedom and divine sovereignty. The Sages' acceptance of astrology was a result of their limited knowledge of mathematics: "Do not ask me to show that everything they [the Sages] have said concerning astronomical matters conforms to the way things really are. For at that time mathematics were imperfect" (Maimonides, twelfth century [1963], part III, chapter 15: 459). Maimonides also resisted literalist interpretations of the Torah; for example, he did not see the six days of creation in Genesis, chapter 1, as a chronological sequence of events. Rather than showing how the universe was formed, Genesis provides insights about the structure of the universe (see Slifkin, 2008 for discussion). The mere frequency of Torah passages is no guideline to how they should be interpreted. For example, there are several scriptural references to God's body parts, yet Maimonides, and most other Jewish theologians, insisted that God is disembodied. Rather, interpreting the Torah should be done in line with our total body of knowledge. If science is clear that a literal interpretation will not work, then the literal reading should be rejected and other interpretations must be sought (Maimonides, twelfth century [1963], part II, chapter 25). Likewise, Gersonides (Rabbi Levi ben Gershon, 1288–1344) eschewed a literalist reading of the Torah. He went as far as to reject *creatio ex nihilo*, instead seeing God as co-eternal with matter and constrained by its properties.

As a result of this legacy, Judaism shows a wide range of positions, from rejection of evolutionary theory in favor of creationism to acceptance of evolution. Jewish theologians tend to accept the transmutation of species, that is, that evolution has occurred and has led to the emergence of species, though some reject natural selection as a mechanism because it undermines the idea of divine design and, by extension, divine providence in human history. For example, Abraham Kook (1865–1935) and Mordecai Kaplan (1881–1983) accepted transmutation but combined it with a progressivist, teleological picture of evolution that rejects natural selection (Cherry, 2003). By contrast, Yeshayahu Leibowitz (1903–1994), while being critical of the modern evolutionary synthesis, embraced a dysteleological picture of nature (see Section 2).

The Hindu reception of evolutionary theory was also varied. During British colonial rule Hindus in India came in contact with evolutionary theory and sought to assimilate it with their religious beliefs. Their responses are as wide-ranging as those of Christian and Jewish authors, ranging from creationist challenges to acceptance. Some authors, such as Dayananda Saraswati (1824–1883), rejected evolutionary theory on scriptural grounds. Saraswati argued for a form of Vedic creationism according to which the Vedic scriptures are infallible and prefigure any later scientific and technological innovations. God created humans in their present form – not as a single couple, but a few hundred thousand strong – and they migrated from Tibet to other parts of the world (Brown, 2012, chapter 10).

Other Hindus endorsed evolutionary theory. For example, Mahendralal Sircar (1833–1904) outlined an evolutionary theism. Sircar accepted common descent and organic evolution and proposed that the mind was a result of gradual evolutionary processes. He pitted this idea against what he termed the "crude doctrine of the transmigration of souls" or reincarnation (cited in Brown, 2012: 63). At the same time, Sircar defended a theistic worldview and conceived of evolution in teleological terms: evolution was God's way to create order out of chaos, with the human mind as its endpoint. Sircar's efforts to integrate evolutionary theory with more traditional Hindu views were part of a broader nationalistic project: he disagreed with colonialist assumptions that Hindus were incapable of scientific thought and believed a deep and sustained engagement with science was necessary for progress and for a nation-state to fully develop. For this reason, he advocated that Hinduism be integrated with evolutionary theory (Chakraborty, 2001).

Other ambitious attempts to integrate Hinduism and evolutionary theory can be found in avataric evolutionism, which holds that ancient Hindu myths of Viṣṇu's ten incarnations prefigured evolutionary theory. For example, Sri Aurobindo (1872–1950) held that God incarnates into the world in successive times, reminiscent of organic evolution. Avatars are mostly associated with Viṣṇu, who is the preserver/sustainer in the Hindu Trimūrti. His avatars descend into the world to preserve dharma and to fight evil, especially when the cosmos is in crisis. Although Viṣṇu's avatars are innumerable, the Garuda Purāṇa lists ten major ones, including a fish, a turtle, a boar, a man (Rama), Kṛṣṇa, and Buddha. Aurobindo proposed a metaphysical picture that saw physical and spiritual evolution as manifestations of God, criticizing Darwinism for focusing too much on self-preservation of organisms at the expense of cooperation: "[B]ecause the struggle for survival, the impulse towards permanence is contradicted by the law of death, the individual life is compelled, and used, to secure permanence rather for its species than for itself; but this it cannot do without the co-operation of others; and the principle of co-operation and mutual help" (Aurobindo, 1914–1918 [2005]: 212). These examples from Judaism and Hinduism demonstrate that the challenge of evolution to religion is not unique to Christianity. Indeed, as we will show in the following sections, the challenge of evolution to religion can be better construed as a challenge of particular aspects of evolutionary theory to specific religious claims, such as those about teleology in nature, human origins, and the origin of religion.

1.3 Why Does Evolution Challenge Religion?

While evolution is not the only scientific theory against which there is religiously motivated resistance, it is the most prominent one. Before evolution through common descent was widely accepted in biology, cosmology was the main battlefield for science and religion, as medieval geocentrism clashed with heliocentrism and later cosmological models. What makes a scientific explanation susceptible to religiously motivated conflict?

One might be tempted to think that any scientific explanation could clash with any religious claim, but in many domains of everyday life people appeal to both natural and supernatural explanations, often integrating both types of explanation. For example, Banerjee and Bloom (2014) found that both theists and atheists have a tendency to ascribe purpose to significant life events such as meeting one's partner, job loss, or illness. So, for example, Carla could say that meeting her future partner Sophie (fortuitously seated next to her on a plane) was "meant to be." At the same time she is aware of the natural causes that put them together: as Carla and Sophie are both religious studies scholars, it is unsurprising they were both flying to the American Academy of Religion's annual meeting from a major international airport. Carla is aware of these naturalistic causes but might still insist that God, or fate, or the universe brought her and Sophie together. Such combinations of supernatural and naturalistic explanations have already been observed by early anthropologists such as E. E. Evans-Pritchard (1937 [1965]), who famously found that the Azande (a tribe in north central Africa) were well aware that termites can cause buildings to collapse. Yet, at the same time, they invoked supernatural agency (witchcraft) to explain why *this* granary collapsed on *that* person. More recently, Cristine Legare and colleagues (2012) found that South Africans attribute AIDS both to natural biological causes (HIV virus) and to supernatural causes (witches) who put one in the way of the HIV virus. It would seem from everyday explanations that natural and supernatural causes are not inherently in competition.

As Legare et al. (2012) observe, there are three domains that consistently generate both natural and supernatural explanations: illness, death, and (human) origins. These three domains have emotional relevance, speaking to events that affect us personally and that are relevant for beliefs about our destiny, both in personal terms (death) and as a species (evolution). For each of these domains, there are well-developed narratives that appeal to supernatural agents and properties, for example, the existence of a soul that lives beyond the death of a physical body. Such narratives predate evolutionary theory and are usually well embedded within cultural contexts. While evolutionary theory makes a wide range of factual claims that seem, prima facie, incompatible with scriptural claims about the age of the Earth, the origin of species, and the position of humanity in the world, a number of authors have argued that religious belief does not require a literalist interpretation of religious origin stories. For example, the theologian Denis Lamoureux (2008) argues that biblical origin stories such as Genesis, chapters 1–3, should not be taken literally. Taken at face value, the Bible has many inaccuracies that even creationists do not accept, such as that there is water in the heavens above the firmament. Instead, the Bible reflects ancient science and accommodates people who read the Bible books at the time of their writing (Lamoureux, 2008: 272). With this approach, it is possible to simultaneously endorse evolutionary theory and be a religious believer. Coming to grips with evolutionary theory, however, might hold some uncomfortable conclusions about our place in the world as just another species (see Section 3). Poling and Evans (2004), for example, show that children and laypeople find it difficult to accept that extinction is inevitable for all species, including humans, while evolutionary biologists endorse the inevitability of extinction of all species, including humans. As evolutionary theory also examines the origins and fate of humanity, it is likely to challenge religious frameworks.

This Element focuses on three challenges of evolution to specific religious claims. The first is metaphysical. Religious worldviews tend to presuppose a teleological understanding of the origins of living things, including human beings, but contemporary evolutionary theory (at least, in a standard sense, as we will qualify later on) understands evolution as nonteleological. The second challenge focuses on human origins: religious and scientific accounts of human origins are not aligned, at least not in a straightforward sense. The third challenge concerns the evolutionary origins of religion itself. Evolutionary explanations of religion, including religious practices and beliefs, may cast doubt on their justification. We demonstrate how these tensions arise and offer potential responses on behalf of some religious traditions. We conclude that it is possible for religions to meet these challenges if some religious metaphysical assumptions are modified.

2 Teleology, Divine Purpose, and Divine Design

2.1 Chance and Evolution

Ancient Greek and Roman philosophers such as Socrates and Cicero often saw the world in creationist, teleological terms (Sedley, 2007). Prior to Charles Darwin, teleological thinking was ubiquitous. For example, the Hindu philosopher Adi Śaṅkara, writing in the first half of the eighth century, thought the world's apparent congeniality to human life provided evidence that it was intelligently designed by a divine creator – the world is suited to human habitation, which is inexplicable through mechanistic processes (Brown, 2008).

This focus on design and determinism in many religious explanations is in tension with stochasticity, a central element of evolutionary theory. A chance or random event seems to occur without any (discernible) cause either because the outcome is genuinely underdetermined or because we lack enough information to make an accurate prediction. Before the rise of probability theory and statistics in the late seventeenth century, stochasticity (chance) was seen as inherently unpredictable. Advances in mathematics and statistics, however, together with collections of data such as census records have made it possible to perceive regularities in phenomena such as games of chance, birth and death statistics, and economic risks. Chance plays a role in evolutionary theory in at least three ways. First, variation, a result of random, unpredictable mutations, provides the raw material with which natural and sexual selection work. Second, genetic drift is the chance disappearance of genes within populations; it can cause major evolutionary changes. Even in the absence of selection, genes can be lost as a result of sampling errors. Third, mass extinctions caused by random events such as asteroids hitting the Earth (which caused the Cretaceous–Paleocene extinction event some 66 million years ago) have played a major role in evolutionary history.

Nevertheless, as we will see in more detail, evolutionary theory did not end progressivist, teleological thinking, as it continues in both scientific and religious writings. For example, the naturalist Ernst Haeckel (1886) saw evolution as progressive and increasing in complexity, with humans at its apex. Jewish authors, such as rabbis Abraham Isaac Kook and Mordecai M. Kaplan, and Christian thinkers, such as botanist Asa Gray (1810–1888) and theologian Frederick Tennant (1866–1957), attempted to integrate evolution within a broader teleological framework (see also Section 3). As we will see, theistic scientists and theologians who seek to incorporate chance and contingency within a broader framework of divine providence are still reasoning teleologically, but their teleology differs from that before the introduction of evolutionary theory.

This section examines whether evolutionary theory challenges teleology, in particular, whether this theory should lead us to think that there is no overall (divine) higher purpose. We first show that teleological thinking is a result of early-developed biases and explore the relationship between this type of thinking and theism. We then review how evolutionary theory challenges teleology, particularly at the macro level. Next, we consider responses from Jewish and Christian authors to this challenge. These responses either reinterpret evolution as teleological or accept stochasticity and argue that it is not incompatible with theism. While the latter type of response is less problematic than the former, it still comes at a cost.

2.2 Intuitive Teleology

A large body of empirical literature suggests that teleological thinking is not merely a product of culture but that it reflects a way human minds make sense of the world. Most of these studies have been conducted with Western children; they show that teleological thinking arises spontaneously. In a typical experiment (e.g., Kelemen, 1999), primary school–aged children are presented with a series of pictures with nonbiological natural kinds such as clouds and rocks, and biological organisms such as tigers. They are asked to choose between explanations for why these things are the way they are, which include teleological explanations, for example, "rocks are pointy so that animals wouldn't sit on them and smash them," as well as causal nonteleological explanations, for example, "rocks are pointy because little bits of stuff piled over a long time." Deborah Kelemen (2004) found that children between about five and ten years of age prefer teleological explanations for all kinds of objects, whereas adults use them only for biological phenomena (e.g., adults endorse that a giraffe has a long neck so that it can reach leaves in high trees).

Subsequent experiments have demonstrated that teleological thinking is a cognitive default to which people resort if they have no alternative explanations. For example, when put under time pressure (commonly referred to as "speeded condition") adults tend to endorse teleological incorrect explanations, for example, "the sun radiates heat because warmth nurtures life," but they still reject false mechanistic explanations, for example, "hills form because floodwater freezes" (Kelemen & Rosset, 2009). Education mitigates teleological thinking. Holding a PhD in the sciences or humanities decreases acceptance of false teleological explanations, although acceptance still increases under speeded conditions (Kelemen et al., 2013). Romani people living in Romania value their traditions, which are transmitted informally at home. As a result only one-third of Romani primary school–aged children regularly go to school.

Casler and Kelemen (2008) found that Romani adults with little formal schooling were more likely to endorse teleological explanations for nonbiological natural kinds, for example, sand is grainy "so that it wouldn't get blown away and scattered by the wind."

Taken together, this research indicates that teleological thinking is a cognitive default stance that arises early in development and that can be mitigated by mechanistic explanations. While it is uncontroversial that teleological thinking is intuitive and overactive (but see Greif et al., 2006), the psychological link between teleological thinking and theism remains unclear. Kelemen (2004) has argued that children are intuitive theists, as young children not only over-attribute teleology but also hold that God is the designer of living and nonliving natural entities. As we have argued earlier, however (De Cruz & De Smedt, 2015), teleology does not automatically entail divine design. Something can be purposive by accident; for example, a tree stump can be suitable as a chair and used as such without having been designed for that purpose.

Current evidence does not show an automatic link between teleological thinking and theism. Lombrozo et al. (2007), for example, found that although Alzheimer's patients were more likely than healthy older controls to endorse teleological explanations for features of their environment (e.g., they were more likely to think that rain exists so that plants and animals have water for drinking and growing), they were not more prone to endorse God as the creator of these features. Recent research also shows that people who are not theists but believe that the Earth is alive, has agency, responds to the needs of animals, and helps them survive (which are termed "Gaia beliefs"[3]) are more likely to endorse teleological explanations. For example, Kelemen et al. (2013) found that although endorsement of teleological explanations among natural scientists was low (e.g., few scientists accept that "germs mutate in order to become drug resistant"), scientists who have stronger Gaia beliefs are more prone to teleology than scientists who hold explicit traditional theistic beliefs or no supernatural beliefs. Cross-cultural research confirms that the link between teleology and theism is tenuous. Järnefelt et al. (2015) showed participants from the United States and Finland images of natural objects such as a giraffe, a maple tree, a mountain, and the paw of a tiger, and then asked them to judge whether this object was purposefully made by any being – the nature of this being was deliberately unspecified. They found that both God-beliefs and Gaia-beliefs were positively correlated with participants' propensity to judge that natural objects were purposively created. In a study with Chinese participants

---

[3] This does not mean that these people entirely endorse the Gaia hypothesis (e.g., Lovelock & Margulis, 1974), but rather, it measures the extent to which they attribute agency and care to Nature or Earth or see her (Mother Nature) as an agent.

Järnefelt et al. (2019) found that, although the majority of participants self-identified as atheists, they all engaged in some ritual behaviors such as veneration of and offering to ancestors, use of lucky charms, and feng shui. The more participants engaged in such practices, the more they endorsed that natural objects were made by a being. This cross-cultural research indicates that there is a link between belief in the supernatural (not just theism) and teleology. This link does not happen automatically, as teleology and divine design do come apart in experiments – not everyone who endorses teleology also endorses the claim that natural objects are purposively designed by a being – but there is still a cognitive association between teleological thinking and belief in the supernatural.

2.3 How Evolution Challenges Teleology

2.3.1 Teleology and Science before Evolutionary Theory

The scientific challenge to teleology predates evolutionary theory. A profound reconceptualization of teleology occurred in the seventeenth century. The Aristotelian concept of teleology holds that each substance has its own final cause. To explain the generation of an organism, such as a budding acorn, one needs to look at its final stage, a mature oak tree. Aristotelian biological explanations are teleological explanations in which the final cause (a mature oak tree) has precedence over the efficient cause (the physical processes through which the tree grows). Aristotle rejected Empedocles' suggestion that we can explain the shape of the human spine as a result of a fetus twisting and turning in the womb. This explanation in efficient causal terms does not explain why the fetus is able to turn and twist in this way. Instead, one should look at a fully developed human and the function of her spine – this explains why the spine develops in an embryo the way it does. Aristotle considered another rival explanation for apparent teleology: perhaps adaptive features might be a result of chance, e.g., it just happens that human front teeth are sharp for cutting food (incisors) and broad at the back for grinding food (molars). He rejected this explanation because all humans have the same types of teeth in the same places, which would be an inexplicable coincidence. Aristotle maintained that final causation is the best explanation: we have to explain what teeth do in order to account for their shape (Falcon, 2015). The scholastics held that each kind of object in the world has its own teleology, expressed in terms of potentialities and liabilities (Boulter, 2019). Efficient causation, in this view, was merely a matter of reducing potentiality to actuality; for example, it is within the nature (potentiality) of iron to rust, and circumstances can bring it about or prevent it from happening. By contrast, it is not within the nature of wood to rust. Things have a natural bias to act in some ways and not in others. These potentialities are within objects, but this happens without reflective awareness or deliberation, except for agents such as humans. As Aquinas (thirteenth century [1998]: 72) observed, "natural agents can tend to goals without deliberating." Aquinas drew on Ibn Sīnā's analogy of the oud player, who plucks the strings of his instrument and thereby creates music without deliberation. If it is possible for a human agent to do things without deliberation, it is also possible for natural agents such as plants and rocks to do so. Although divine action was important for the scholastics, this picture of teleology required no immediate divine intervention. Teleology made sense entirely in terms of secondary causation, that is, causation by created things.

The early modern mechanistic conception of nature as governed by natural laws overturned this Aristotelian picture. Authors such as Descartes, Boyle, and Newton devised a legalistic framework to think about nature: rather than conceiving of liabilities and powers of individual kinds of things, all objects in the world obey the same general principles (Jaeger, 2008). Nature obeys God's unbending and absolute laws, which can only be broken by God when performing miracles. Where did this leave teleology? As an unexpected result of this mechanistic worldview, the idea of final causes became no longer intelligible: if things don't have an intrinsic nature and teleology but only obey some general laws of nature, how do we make sense of adaptive design in nature? One solution to this puzzle, offered by natural theologians, was to explain teleology as the result of divine design. Adaptive features of organisms, such as the eye, the shape of teeth, and the swimming bladder, were marvelous contrivances that showed the ingenuity of the creator. Even astronomical features, such as the number of planets and their distance to the sun, were explained as products of divine providence. During the seventeenth century the design argument for God's existence enjoyed its heyday (McGrath, 2011). Authors such as Ray and Nieuwentyt took the complexity, ingenuity, and goal-directedness of features of the natural world as evidence for intelligent design. In the absence of final causation, an external cause was invoked to explain why the eye was so well designed and suited for seeing. This cause was the divine legislator, who had created the natural laws. Natural theology continued well into the nineteenth century, although philosophers such as Hume and Kant had criticized it (Ruse, 2003). For instance, Immanuel Kant (1790 [2000], §68) charged the design argument with vicious circularity, as God helps natural scientists to make the finality of nature explicable, while at the same time "we turn round and use this finality for the purpose of proving that there is a God." The design argument thus had an important epistemic function, making sense of apparent teleology in a world of laws and purposeless matter.

In the final decades of the seventeenth century, and especially during the eighteenth century, three distinct traditions arose that explained teleology in naturalistic terms: evolution, epigenesis, and transmutation. (Note that the terms evolution and epigenesis had quite a different meaning then than they have today.) Evolution and epigenesis were two competing theories that explained how an embryo grows into an organism without reference to final causes (Richards, 2000). The theory of *evolution*, mainly developed by the entomologist Jan Swammerdam in his *Historia insectorum generalis* (1669), held that embryos grow because female gametes already contain the adult form, with the semen only acting as a stimulus to realize the adult type. Like other evolutionists, Swammerdam had a place for God in his picture: humans are tainted by original sin because they were literally encapsulated within Eve's eggs (an intriguing female variation on Augustine's male seeds within seeds, see Section 3.5). The rivaling theory of *epigenesis* stated that embryos started out formless and only gradually took on their definite form. Without final causes, however, some sort of mechanism had to be postulated to explain why human embryos grow into humans, whereas dog embryos grow into dogs. The physician Caspar Wolff postulated an essential force to explain how this occurs in his *Theoria generationis* (1759). In his short treatise *Über den Bildungstrieb und das Zeugungsgeschäfte* (1781), the physician and physiologist Johann Friedrich Blumenbach proposed a single unified force (*Bildungstrieb*) to explain how wounds heal, polyps regenerate, and embryos grow. The *Bildungstrieb* was a naturalistic, but teleological force. *Transmutationist* theories, one of the direct precursors of evolutionary theory as we know it today, argued that species could change (transmute) into different species. An early example, *Telliamed* (1748), proposed that all life originated in the sea and that marine species transmuted into land organisms; for example, seals changed into dogs and seaweed developed into shrubs.[4] In spite of the radicalness of *Telliamed*, transmutation became increasingly popular across Europe. But it left unexplained the inheritance of traits, which epigenesis could arguably explain better at the time by positing the *Bildungstrieb*. It would take until well into the twentieth century for evolutionary theory and Mendelian genetics to be synthesized into a single theory (the modern synthesis) that explains both apparent teleology and inheritance.

Of these three naturalistic theories, transmutation became increasingly popular. Charles Darwin's grandfather, Erasmus Darwin, was a proponent, as were

---

[4] Benoît de Maillet was worried about repercussions of this theory, hence he published it anonymously (the title is his name in reverse) and posthumously and had the book printed in "Amsterdam." Many French books printed illegally, without the obligatory royal privilege, were labeled as printed in Amsterdam, but were actually published in Paris by rogue printers.

many other scholars in the nineteenth century. Evolutionary theory did add, over time, many novel elements that were not found in evolutionism, epigenetics, or transmutation. One of its distinctive features, especially since the modern synthesis, is its ability to provide an epistemically satisfying naturalistic solution to the question of how apparently goal-directed features, such as eyes, wings, and hands, and ecological adaptedness can arise in the absence of Aristotelian natures (which provide their own teleology) and without having to appeal to a divine designer or mysterious formative forces. But the broader metaphysical ramifications of evolutionary theory and the place of teleology within this naturalistic picture remain the topic of continued scientific and philosophical debate. We will now examine to what extent evolutionary explanations can still be construed as teleological in some sense before considering the ramifications for religious and, in particular, creationist explanations (in the broad sense of explanations that involve creator gods).

2.3.2 Is Teleology Appropriate in Biology in the Light of Evolution?

There is continued debate among philosophers and biologists on whether teleological explanations are appropriate in biology (e.g., Ruse, 2016). To take an example that was already discussed by Darwin and his contemporaries, orchids have a peculiar shape whereby the pollinia are separated from the stigma, the female part of the flower that receives the pollen, by a flap of tissue, the rostellum. This makes self-pollination unlikely and increases the probability of cross-pollination. Cross-pollination is beneficial for the resulting offspring as they are more genetically diverse and thus less susceptible to disease. Some short-blooming genera, such as *Sobralia*, rely on temperature signals to coordinate their blooming, thereby increasing the probability that cross-pollination will occur. Orchids also have exclusive relationships with moths, hummingbirds, and other pollinators. Orchids of the genera *Ophrys* and *Cryptostylis* lure their pollinators by mimicking the appearance of sexually receptive females, deluding males into approaching and pollinating the flowers while never providing a reward in terms of nectar. When considering these peculiar adaptive features of orchids, can we say that orchids have these particular structures so as to encourage cross-pollination? This explanation appears to be future-oriented: the orchid develops the appearance of a sexually receptive female so as to tempt male pollinators to visit it. This future state of affairs may never happen since, for example, the orchid might be eaten by aphids before it reproduces. In a non-Aristotelian universe, such future-orientedness is mysterious. Neoteleologists (e.g., Neander, 1991) argue that teleological explanations are appropriate: we can indeed say that the rostellum of an orchid is there so that it would not self-pollinate, or that an *Ophrys* has its particular shape so that it can be cross-pollinated. But neoteleologists deny that teleology is forward-looking. Rather, a given trait has its function because of the history of that trait, that is, the evolutionary function it had in the past. If a trait fulfils this function, it is properly functioning. Karen Neander (1991) gives the example of near-sightedness in penguins. When on land, penguins are myopic due to the curvature of their eye lenses. But this curvature allows them to see clearly under water, which is vital for their aquatic hunting. The myopic curvature of the penguin eye can be explained as a result of its evolutionary history, namely selective pressures on the eye that made it more suitable for underwater hunting. By contrast, near-sighted humans do not have properly functioning eyes; there is nothing in their evolutionary history that suggests this would be advantageous. The notion of function is thus etiological: it looks at the evolutionary history of a particular trait (e.g., penguin eyes) to explain particular functional features (e.g., being able to hunt under water).

The assumption that the current function of a trait is a result of its evolutionary history remains contentious. Robert Cummins (2002) has objected to this claim, arguing that we don't need to appeal to historical factors to explain current function. For example, the reason birds have wings is not the same evolutionary reason that the first, rudimentary dinosaur wings evolved – these primitive wings did not enable full flight yet, but may have assisted in capturing small prey, leaping into the air, or incline running. A second problem with neoteleology is that it looks at function in an essentialist way that provides little scope for developmental plasticity. A trait may be able to function atypically, even though it does not work according to some imagined blueprint – disabled animals do get by (Amundson, 2000).

Even if we grant that there may be teleology in nature, naturalistically understood as the result of blind evolutionary processes, it is harder to reconcile evolution and teleology at the macro level. For example, Ernst Mayr (1992) accepted the idea of teleology (which he termed "teleonomic processes") in particular organisms, but ultimately thought these processes were reducible to material (physical, chemical) processes. By contrast, he held that the idea of a global or cosmic teleology would necessarily require some supernatural elements. For this reason, Mayr rejected cosmic teleology. Similarly, Jacques Monod accepted that teleology exists within biological organisms at the level of adaptive complexity, but he advocated that living beings are ultimately the result of chance. To see a larger teleological picture was, in his view, a fundamental mistake. Instead, Monod recommended that people come to grips with their position as creatures in a universe that is ultimately indifferent to them:

If he accepts this message in its entire significance, then man must at last wake up from his millenary dream to discover his total solitude, his radical foreignness. He knows now that, like a wanderer, he is in the margin of the universe where he must live. A universe that is deaf to his music, as indifferent to his hopes as it is to his suffering or his crimes. (Monod, 1970: 187–188, our translation)

Nevertheless, some contemporary philosophers want to restore teleology, or at least directionality at the macro level, using concepts such as niche construction and eco-engineering. These evolutionary views attempt to reinstitute the importance of agents; they also allow for naturalistic macroteleological explanations. As Samir Okasha (2018) notes, biological entities such as beavers are often conceptualized in terms of agents who pursue goals; sometimes the evolutionary process itself is conceptualized as an agent who "optimizes" design and selects the fittest organisms (Mother Nature as agent). For example, Sewall Wright and other mid-twentieth-century evolutionary biologists argued that natural selection led to changes in gene frequency so that mean population fitness was always maximized. This form of reasoning is problematic, however, as natural selection is not an optimizing process in this strong sense (Okasha 2018, 20). We cannot simply replace a designing God with a designing process of natural selection and restore teleology at the macro level.

A second way in which global teleological trends can be discerned in nature without personifying natural selection or nature focuses on organisms as agents. For example, Peter Godfrey-Smith (2017) discerns directionality toward more complexity in the history of life as a result of the active role organisms play as subjects and causes of their own and other creatures' evolutionary trajectories. He argues against the traditional evolutionary picture, which conceives of the evolutionary history of a species as a passive one-directional process that sees a species as a plaything of the environment and its vagaries, leading to the selection of genes that code for phenotypic traits that fit the environment. By contrast, in Godfrey-Smith's naturalistic picture, there is no overall "designer" (such as natural selection), but we can still discern teleological trends at the macro level because of the role that subjects play in directing the evolutionary paths of their descendants and other species, for example, the role that predators play in shaping the evolution of their offspring and the species they prey upon, or the active role that organisms can play as causes. Organisms such as beavers actively shape their environment and that of other species living in it (beaver impoundments drastically change the local ecosystem), and thereby influence the course of evolutionary history.

2.3.3 Contingency or Convergence?

To get a better sense of whether there is some form of directionality to evolution, we need to evaluate macroevolutionary patterns more carefully. Two approaches consider the role of chance in macroevolution: contingency and convergence. The main proponent of *contingency* is Stephen Jay Gould (1989), who stated that if the tape of life were rewound to the time of the Burgess Shale (ca. 500 million years ago) and played anew, we would likely not end up with the life forms we see today. The Burgess Shale is a fossil deposit from the Middle Cambrian, showing many exceptionally well-preserved fossils, including many morphological forms that have since disappeared. Why did, for instance, *Pikaia*, a probable ancestor to the chordates (including humans), leave descendants, whereas so many other organisms did not? Why was this small (less than 4 cm long) leaf-shaped creature with its antennae, tail fin, and eel-like movement so successful? According to Gould, *Pikaia*'s success and the extinction of other Cambrian species were not consequences of natural selection. Rather, the long-term survival of lineages is akin to a lottery that could easily have turned out otherwise.

Extinction is often a case of bad luck, not just bad genes (Raup, 1991). For this reason, it is impossible to predict which species will go extinct. Highly successful species regularly go extinct due to exceptional factors beyond their experience. For example, Neanderthals were long considered as somehow inferior in intellect to humans and their demise was seen as a result of competition with superior Aurignacians (the earliest *Homo sapiens* population that colonized Europe). It has become clear, however, that Neanderthals were highly successful and culturally sophisticated top predators who fell victim to climate change (Finlayson, 2009). There are only a few empirical studies that tackle contingency as a testable hypothesis. For example, studies by Jonathan Losos and colleagues (reviewed in Beatty, 2006) on the evolution of lizards living on four Caribbean islands can demonstrate *local* instances of contingent evolution, but do not give us a sense of whether Gould's claim about *global* evolution is true.

The second framework that looks at evolution and potential directionality is *convergence*, defended by Conway Morris (2003) and McGhee (2011). Convergence is the claim that evolution often produces similar solutions to similar problems, so we end up with recurring adaptations and analogous ways in which organisms fill niches. Simon Conway Morris (2003) makes the strong claim that once particular initial conditions are set, certain biological adaptations are inevitable: given a universe like ours, humans will inevitably emerge. George McGhee (2011) offers a more systematic argument for convergence that covers a catalogue of convergences in animals, plants, and ecosystems. For example, spikes on leaves as a defense against being eaten have evolved in seven clades of plants, and spines on twigs or bark have convergently evolved nine times. In animals, paddle-shaped swimming appendages have emerged seven times independently in clades as distinct as sea turtles, diving beetles, plesiosaurs, penguins, and manatees. Eyes have convergently evolved 49 times. Deep homology poses a potential problem for convergence. For example, the highly conserved pax-6 gene underlies the evolution of camera eyes in vertebrates and compound eyes in arthropods. But McGhee (2011) is not too concerned with this. He sees deep homology as an instance of convergence: the fact that pax-6 helps regulate eye development is still due to a selective pressure that is ubiquitous on our planet, namely the Sun as the main source of light.

While contingency and convergence are scientifically informed hypotheses, they paint also broader, ontological pictures that stress aspects of nature (chance for contingency, necessity for convergence) that evolutionary theory deals with. For each it is possible to study local patterns, but the global picture remains elusive. A problem for convergence is that it looks at local instances of adaptation and gerrymanders them into a global teleology. There are instances of convergence but no instances where convergence failed to materialize. Since it is difficult to establish a base rate, it is difficult to assess whether evolutionary convergence is inevitable, as Conway Morris holds. The lack of a base rate can also be put on the doorstep of the proponents of contingency. At the moment, we know of only one planet where life has evolved and where we can look at patterns of evolution, so the theory remains difficult to test. Both frameworks are clearly onto something: the life forms we ended up with are the result of historically contingent processes but also show some convergence.

2.4 Chance, Determinism, and Theism

As we saw above, contemporary evolutionary biologists have created ontological pictures of life on Earth by stressing either contingency or convergence. Evolution still seems to leave some room for teleology, but it remains unclear whether teleological pictures at a macro level can be invoked in a naturalistic framework. Theistic positions as developed in Christianity and Judaism, for instance, do propose global teleology in the form of divine providence, God's relationship with the world and God's actions in it. We will review three such positions here, all of which propose some guise of creationism (i.e., a creator God) as starting point and look at contingency in nature to draw conclusions about divine action: stochasticity/contingency as a mere feature of how we perceive the world; stochasticity as a genuine feature of the world, but with God in charge; and the world as both stochastic and unknowable (even to God).

2.4.1 A World That Is Only Apparently Stochastic

Maybe evolution isn't truly unpredictable – it just appears that way to us. David Bartholomew (2008, 189–190) suggests that divine goal-directed action – in particular divine design, the idea that God creates and shapes creation in a purposeful way – does not commit one to convergence. Divine design, down to the minutest details, is perfectly compatible with a Gouldian outlook of contingency, albeit not with its ontological and epistemological ramifications. Keith Ward (1996) likewise holds that not only is it appropriate for a theist to believe that evolution is guided, but the evolution of intelligent life (humanity) is more probable given theism than atheism: if the life forms we see today resulted from evolution without divine intervention, then intelligent life is highly improbable, but assuming theism, it is highly probable. Alvin Plantinga (2011) holds that evolution and theism are compatible but rejects the idea of unguided natural selection: it is perfectly possible that "the process of natural selection has been guided and superintended by God, and that it could not have produced our living world without that guidance" (p. 39). According to Plantinga, mutations are not truly random, as God could "guide the course of evolutionary history by causing the right mutations to arise at the right time and preserving the forms of life that lead to the results he intends" (p. 121). Curiously, Plantinga hereby harkens back to a premodern concept of chance, which saw chance events as vehicles for divine revelation, as exemplified in the use of chance outcomes in ancient oracles (Reeves, 2015). For authors from late antiquity, such as Augustine (fifth century [1972]), things appear random because God's purposes are hidden from us. In this view the stochasticity governing evolution and other natural processes is not an ontological feature of the world.

Plantinga presents naturalism and theism as a binary choice. Fiona Ellis (2014) challenges this dichotomy, arguing instead for an expansive naturalism that can include theistic naturalism. She agrees with the philosophical orthodox position of naturalism that we are natural beings, living in a natural world, and that our claims should be empirically grounded, but she rejects the assumption that the only objects naturalists can study should be those that scientists happen to take an interest in. One can be a naturalist who accepts evidence of natural selection and yet also be a theist. So perhaps authors like Plantinga force a false choice between naturalism and theism.

Plantinga's and Ward's position that stochasticity is only apparent comes at a cost. It exacerbates the problem of evil for the same reason that intelligent design creationism does. Take Behe's (1996) claims about how the malaria parasite is ingenuously created and how its resistance to chloroquine is inexplicable through natural selection alone. Given the devastation that malaria causes, one may ask why an intelligent designer would allow something as destructive as malaria in the first place. In a similar vein, one can ask why God, masterminding mutations in the right place at the right time, would not prevent the emergence of harmful mutations that create suffering. For example, God would be directly responsible for every instance of Tay-Sachs disease, a rare but fatal genetic disorder in fetuses and young children that gradually destroys the nervous system, leading to death by age five.

In a letter to Asa Gray, Darwin (1860) reflected on the *Ichneumonidae*, a parasitoid family of wasps that insert eggs into the bodies of living caterpillars, and rejected direct divine intervention or guidance of evolution because of the cruelty of the wasp larvae eating their living hosts from the inside: "I am inclined to look at everything as resulting from designed laws, with the details, whether good or bad, left to the working out of what we may call chance." But even this was ultimately unsatisfying to Darwin. Incorporating evolution and its stochasticity into a theistic framework has two elements that seem to pull in different directions. On the one hand, evolution results in a lot of suffering through competition, predation, and harmful mutations. On the other hand, evolution has resulted in intelligent life, particularly humans, to which many religious traditions accord a special place.

2.4.2 A Stochastic World Where God Is (at Least Partly) in Control

A second position holds that stochasticity is a genuine feature of the world, but that God is (at least partly) in control. Maimonides distinguished between general providence, which has a role for stochasticity, and special providence, which hasn't. Although he was obviously not aware of evolutionary theory, his reflections on the reach of chance and providence (Maimonides, twelfth century [1963], part III, chapter 17) are informative for the present discussion. Maimonides considered, and rejected, the Aristotelian picture of teleology, the atomist picture (everything results from the chance collision of atoms), and two Muslim philosophical schools, the Ashʿarites (according to whom nothing is up to chance, as God exerts equal causal force on everything) and the Muʿtazila (who rejected chance as God gives everyone their due). The latter would claim that God controls all events, even "the falling of this particular leaf." But that means that we need to ask, "because of what sin has this particular animal been killed?" (Maimonides, twelfth century [1963], part III, chapter 17: 468) The Muʿtazila believed that God would provide compensation for that animal in the afterlife. Some contemporary Christian theologians, such as Southgate (2008, see also below), defend a similar compensation theory for nonhuman animals. Maimonides found this rather far-fetched. Instead, he advocated an intermediate position, which allows for genuine chance in nature. For example, it is not the case that "when this fish snatched this worm from the face of the water, this happened in virtue of a divine volition concerning individuals," but rather that such instances are "due to pure chance" (Maimonides, twelfth century [1963], part III, chapter 17: 471). Divine providence is graded: God looks out directly for elected morally upstanding humans (special providence), while leaving animals and other humans, who are not in God's favor, subject to the vagaries of life (general providence). God's special providence grants wisdom to the elect as only rational animals are in a position to receive ideas and insights from God, for example, prompting them to make wise decisions. Because they lack rationality, nonhuman animals could be subject only to God's general providence and so are subject to chance events. Other Jewish theologians from this period agree; for example, Gersonides accepted divine special providence for the saintly only.

In this position, God is in control, but stochasticity is a genuine feature of the world (and thus part of general providence). God carefully chooses to exert divine agency and loads the dice where God wants, leaving the rest up to chance. How then to explain scriptural passages that hint at providential care for animals, such as "The young lions roar for their prey, seeking their food from God" (Psalm 104:21)?[5] Maimonides responds that providence works at the species level, but not for individual animals, a position Jerome Gellman (2009) calls holistic providence. Gellman compares holistic providence to convergence in contemporary biology, suggesting that God would impose boundary conditions and constraints to give a rough outline of how evolution would go – convergence across species would be due to God's providential work. While this position does a good job in maintaining both chance and providence as features of the world, it faces difficulties. Is holistic providence enough for God to be just toward creation? Animals (including humans) frequently suffer, and the idea of divine compensation for endured suffering might not be as ludicrous as Maimonides supposed – without compensation, such suffering seems unjust. John Schneider (in press) argues that God maximizes beauty in creation and uses evolution to achieve it, but he also holds that without compensation for the suffering creatures, it would be morally monstrous. Therefore, compensation

---

[5] All Bible verses are from the New Revised Standard Version (NRSV).

should be part of any eschatology that explains evolution. Next to this, a clear distinction between rational and nonrational animals is implausible in the light of comparative psychological studies – particularly on animal rational choice and cognitive flexibility, including social and numerical cognition and foraging – that show that the gap between humans and other animals is one of degree rather than kind.

Contemporary neo-Thomism also sees genuine chance in evolution and other natural processes, yet accords a role for divine providence. This is not a graded model of providence, but one that invokes two levels of causation (e.g., Johnson, 1996). Neo-Thomists, following Aquinas, draw a distinction between God as primary cause and secondary causes (natural causes), which are ultimately dependent on God. Secondary causes are complete and have genuine integrity and causal force within the natural order. At the same time, God's causal agency is also complete, but God is not a cause like the secondary causes. This removes the need for God to tinker in nature's gaps or margins, such as in quantum indeterminacy events, as some authors working in science and religion have proposed (e.g., Murphy, 1995; Russell, 2013). Elizabeth Johnson (1996) argues that through contingency God grants the world its own structural integrity: God improvises with the evolving creation, not unlike jazz musicians or dancers who adapt their artistic work to the cooperation of other members of the band or troupe.

The neo-Thomistic model faces some challenges. How to explain this double agency, where somehow both God *and* natural causes jointly (but at different levels) cause the same events to happen? Johnson (1996) ultimately appeals to mystery, which is not very satisfying. Another problem for neo-Thomism is that it seems too flexible: no matter the details of the laws of nature, neo-Thomists can readily account for divine action through their distinction between primary and secondary causation (Ritchie, 2017). But one worries that neo-Thomists display an insouciance about scientific knowledge as they loftily glide over the nitty-gritty details; the specifics of how mutations and macroevolutionary changes happen do not seem to matter for the account. Historically, Aquinas and other scholastics (following Aristotle) saw teleology within each natural kind, due to its potentialities and liabilities. This picture changed over time, with first a mechanistic picture of natural laws with God as legislator in the early modern period, and from the nineteenth century onward, teleology within specific adaptations as a result of natural selection. The neo-Thomist position can readily be applied to these without any modifications but was actually devised to deal with separate teleologies in natural kinds. Ever since it has been marooned from its historical framework, it no longer has a mechanism that could somehow connect the two levels of causation. Without a clear sense of how they connect, neo-Thomistic authors need to resort to analogy. Like Johnson (1996), Stephen Barr (2009) uses creative analogies to explain the joint causation; for example, Shakespeare is the first cause of Polonius's death in *Hamlet*, as he wrote the play, and the secondary cause is a dagger. While such analogies are intriguing, it still seems unclear how we can understand their operation in the natural world, which, at least to us, does not resemble a piece of music or a play.

One advantage of Johnson's model is the emphasis on freedom that evolution allows. Other Christian models propose a more circumscribed freedom. The molecular biologist Kenneth Miller (1999 [2007]: 233–234) agrees that by denying chance, as in a world that is only apparently stochastic (the Plantinga position), God becomes directly responsible for every instance of evil in the world, including "the school bus full of children slipping off an icy road." Miller offers an only-way theodicy: stochasticity is "the only way in which a truly independent physical reality can exist" (created by God). If we assume God created the world, and God also desires creation to be free, evolution through natural selection is, in Miller's (1999 [2007]: 269) view, exactly what we would expect: "a Deity determined to establish a world that was truly independent of His whims, a world in which intelligent creatures would face authentic choices between good and evil, would have to fashion a distinct, material reality and then let His creation run."

Christopher Southgate (2008) endorses a similar theodicy, focusing on the problem of animal suffering that is caused by natural selection. If God chooses natural selection as a mode of creation, and the ultimate end of creation will be a renewed heaven and earth (as most Christian traditions hold), why did God not create this final state right away? The poignant example of pelican chicks illustrates his point that natural selection inevitably results in the suffering of sentient creatures. Pelicans hatch two chicks per clutch, one they invest in and a second "insurance chick" that could be raised if something went wrong with the first one. But typically, the "insurance chick" is ignored by its parents, and its sibling pushes it to the edge of the nest where it dies of starvation. Such suffering cannot be explained by human sinfulness but is a feature of how natural selection operates. Southgate (2008: 48) argues that to make sense of this suffering, it must be the case that evolution is "the only, or at least the best, process by which creaturely values of beauty, diversity, and sophistication could arise." He presents an only-way theodicy where evolution is the only way to create individual "creaturely selves" (not only humans, but also other organisms). While this theodicy avoids problems that many other theodicies have faced, such as Augustine's that blames death on human sin, it still faces a number of difficulties. The only-way theodicy assumes that if God had another way of creating creaturely selves without such a wasteful process that causes suffering, God would have done so. An omnipotent being would not inflict suffering on the world without good reason. But how can Southgate be confident that this is the case? As Holmes Rolston (2018) remarks, there are many beautiful, valuable things in the universe, including galaxies, crystals, and molecules, in which Darwinian processes do not play a role (although stochasticity *does* play a role in their formation). It is not because theologians cannot come up with alternative mechanisms that none are available to God. Southgate further stipulates that not only species but also individual creatures that failed to flourish during their lifetimes, such as the short-lived pelican "insurance chicks," will be redeemed in heaven. This solution poses further challenges, however, such as whether predators in heaven will still chase prey (Southgate thinks they will), or whether lions will be eating straw (Isaiah 65:25).

2.4.3 A Stochastic, Unknowable World

Few theistic authors have embraced a Gouldian contingency with all its metaphysical implications. At first blush it seems hard to reconcile this view with theism. For example, the herpetologist John Reiss (2009: 356) concludes that such a worldview utterly undermines any ideas about divine design: "Life is not designed, or at least it shows no evidence of design for anything other than continued existence, which needs no designer ... We must admit that there is not only not design but indeed not even 'apparent design' in the biological world, in the sense of entities doing any more than they need to do to continue to exist." However, the complexity researcher Stuart Kauffman (2008: 6–7) has defended a position where this lack of predictability is still compatible with theism. He considers God as a human invention but also as a shared sacred space. The unpredictability of natural selection is liberating and creative, and God is not, in Kauffman's view, the Anselmian omni-God but rather someone "who [does] not know or control what thereafter occurs in the universe. Such a view is not utterly different from one in which God is our honored name for the creativity in the natural universe." This is not a creator God but creativity itself. It is difficult to reconcile this view of stochasticity with classic Anselmian theism. Many philosophers of religion and theologians present a dichotomous choice between not just naturalism and theism, but between naturalism and Anselmian omnitheism. This emphasis on the Anselmian omni-God is not particularly helpful when looking at evolution because it precludes a number of plausible alternative options such as Kauffman's.

Another proponent of a radically stochastic view was the Jewish theologian Yeshayahu Leibowitz (1903–1994). Leibowitz's views can be situated in a broader metaphysical project: he rejected traditional religious concepts such as providence or redemption, and separated religious practices from these metaphysical ideas. Leibowitz resisted the idea that following the Jewish law (*halakhah*) would provide any this-worldly benefits. Rather, he saw religious belief as a personal, autonomous decision that says more about the religious believer than about God, who is inscrutable. Leibowitz advocated a separation between science and religion. Unlike Gould, he did not draw a distinction between scientific facts and religious values, but rather, between scientific facts and religious practices. Leibowitz thus did not use evolutionary theory to support his views on Judaism; rather, his views on evolutionary theory fit within a larger picture on how science and religion should relate (Cherry, 2003). Evolutionary theory cannot demonstrate teleology but can only show efficient causation (i.e., not final causation). At the same time, drawing metaphysical conclusions about ultimate reality based on evolution, such as Reiss (2009) does, was, to Leibowitz, overreaching. Drawing on the Book of Job, Leibowitz argued that the divine speeches, where God responds to Job's questions about why he must suffer so much, point to our profound lack of being able to discern any teleology in nature:

> The response ascribed to God by the author of the Book of Job is . . . at first sight, irrelevant. It does not answer the question "whence shall wisdom be found." It does not discuss man and his fate. . . . It leaves physical reality unaccounted for. It describes being as it is without judging it. It presents the cosmic and terrestrial world, from the inanimate to the living, from the splendid and wondrous to the awful and monstrous – especially the monstrous phenomena – without hinting at any purpose in this amazing creation, or any secret intention underlying the monstrosity. Such is the Creator's Providence, and this is what satisfies Job. (Leibowitz, 1992: 52)

God does not provide any explanation for the problem of evil but instead describes creation as it is, including behemoths, leviathans, and ostriches. As a result, "this [divine] providence manifests itself in natural reality itself; It is not intended or directed toward any specific purpose or goal, not even toward the needs and wishes of Job" (p. 52). Leibowitz's interpretation has parallels with contemporary Job scholarship that analyzes the text in its ancient Near Eastern context. For example, Carol Newsom (2003) sees the Book of Job as a polyphonic text that incorporates diverging theological ideas, exemplified in Job, his friends who come to comfort him, and the divine speeches. She interprets the divine speeches as sublime rhetoric. They evoke the sublime by subverting well-established Near Eastern imagery such as the sea being in opposition to God by presenting God as a midwife to the sea's birth, carefully swaddling it in darkness and cloud like an infant (Job 38:8–11). The divine speeches also indicate that God cares for and delights in predators such as hawks, lions, and eagles, and creatures that represent chaos such as ostriches and Leviathan. The chaotic sublime is presented as an integral part of creation (see also Schneider, in press). What the divine speeches indicate is that God does not disvalue chaos, or that God does not even see it as a regrettable but necessary way to create living beings (as Southgate's only-way theodicy holds). Rather, God values chaos, unpredictability, wildness, and predation, all sublime properties, as integral parts of creation.

This third position has the advantage of taking the metaphysical implications of evolution on board (which Plantinga and Southgate see as a problem that the latter attempts to solve with compensation in the afterlife), namely by embracing a lack of teleology in nature. One potential worry with this view is that it makes God either very different from the traditional monotheistic omni-God (Kauffman) or profoundly inscrutable (Leibowitz).

To sum up, two developments profoundly affected the picture of teleology in the seventeenth to nineteenth centuries: the crumbling of the Aristotelian model and the introduction of chance as a real phenomenon that can be studied scientifically. The theory of evolution still leaves open several possible metaphysical interpretations about the role of chance: convergence emphasizes solutions evolution repeatedly comes up with, whereas contingency stresses the unpredictability of evolutionary processes. These two metaphysical interpretations have been combined with theistic views that try to harmonize evolution and religion. As we have seen, each of these comes at a cost. They exacerbate the problem of evil, appeal to mysterious forms of causation, are radically different from more traditional theistic views, or make God inscrutable.

3 Human Origins: An Evolutionary Challenge to Religion?

3.1 Scientific Explanations and Human Origins

Scientific and religious accounts of human origins frequently venture on the same terrain, not only seeking to elucidate human origins but also aiming to clarify our place in nature. Take the paleoanthropologist Clive Finlayson's (2009) detailed overview of what is known about the extinction of *Homo neanderthalensis*. Presenting the case that Neanderthals did not go extinct because they were unintelligent but rather due to climatic events beyond their control, he muses:

It might so easily have gone another way: a slight change of fortunes and the descendants of the Neanderthals would today be debating the demise of other people that lived long ago. This is not a trivial question. Behind it lies the implication that we are not as unique and special as we might think. We owe our existence to a series of events in which chance played a huge role. (Finlayson, 2009: 2–3)

Given the enduring curiosity for our own origins, it is perhaps unsurprising that paleoanthropology is a prestigious scientific discipline, with human fossils frequently making the cover of scientific journals such as *Nature* and *Science*. There is not nearly as much attention for other animal fossils, "no suid [or other nonhominin] skulls grace the covers of *Nature* or garner headlines like 'New pig skull completely overturns all previous theories of pig evolution'," as the paleoanthropologist Tim White (1995, 369) quipped.

Evolutionary theory, combined with disciplines such as geochronology, archaeology, geology, paleoanthropology, and paleogenomics, offers a rich account of human origins. This account is still patchy. However, scientists have moved beyond a search for an alleged "missing link" between humans and other great apes to a detailed archaeological record of ancestral hominins, their tools, their habitats, fine-grained dating of evolutionary events such as the emergence of bipedalism and multiple exits out of Africa, and intriguing genetic clues about population size and interbreeding between different hominin species. This picture reveals that climate change and other chance events played a crucial role in human evolution. It cannot be conceptualized as a single ancestral line, as exemplified by popular (but erroneous) imagery of a chimpanzee knucklewalking in and a *Homo sapiens* striding out in triumph. Instead, human evolution since the split between the last common ancestor of humans and chimpanzees is a messy bush with many dead ends: a mere 60,000 years ago, the world was populated with several hominin species, including *Homo sapiens*, *H. neanderthalensis*, the diminutive Southeast Asian *H. floresiensis*, the elusive Denisovans in Siberia, and possibly late *H. erectus* in Indonesia. Today, only one hominin species, *H. sapiens*, survives.

This section examines scientific views on human evolution and the challenge they present to religion, in particular to Christianity. We focus on the doctrine of original sin, a key element of Christian theological anthropology. Since the doctrine of original sin is at odds with scientific accounts of human origins, some theologians and scientists have attempted to reconcile them. We first examine early attempts at harmonizing scientific and religious views on human origins by Tennant and Broom, and then move on to contemporary accounts of the Fall and attempts to reconcile it with evolution, looking at

Augustinian and Irenaean perspectives. Finally, we consider whether a nonlapsarian view based on Schleiermacher can help harmonize the Fall narrative with human evolution.

3.2 Scientific Accounts of Human Evolution

3.2.1 Early Scientific Work on Human Evolution

According to a familiar narrative, Europeans largely adhered to a literalist, creationist account of human origins that conceived of humans as created in their present form roughly six thousand years ago. Darwin, according to this narrative, challenged this biblical literalism, replacing it with an evolutionary story of human origins, which is now universally accepted among scientists. However, as we will outline below, this narrative is incorrect: biblical literalism was *not* the default stance, but rather is a recent movement that only firmly took hold in the late nineteenth and early twentieth centuries. It actually developed as a reaction to evolutionary theory. Nonliteralist interpretations of human origins were already developed by early Church Fathers such as Origen (184–253) and Augustine (354–430). Origen argued that humans became embodied creatures as a result of the Fall and were spiritual beings before that time. Augustine did not interpret the six days of creation as history, as some contemporary creationists do, although he did interpret other aspects of the creation narrative literally (such as a single ancestral pair as the ancestors of all humanity).

By the eighteenth century, it had become evident that humans and nonhuman primates bear striking physical similarities. For this reason, Linnaeus classified humans and apes together in the order *Anthropomorpha*, later called *Primates*, but he did not propose an evolutionary link between them. Inspired by early indications that sea creatures once lived atop what are now landmasses (in particular fossilized shark teeth), the natural historian Benoît de Maillet (1748) proposed that an ocean once covered the entire planet. Once it retreated and landmasses appeared, sea animals transmuted (evolved) into animals that acquired the ability to walk on land. This work was very influential, both in fiction, giving rise to early French science fiction, and in natural histories. The French natural historian Lamarck (1809) was the first to propose apes as ancestors to humans. He drew on earlier ideas that were widespread among natural philosophers, including the concepts of a vital life force and the transmission of acquired characteristics. The anonymously published *Vestiges of the Natural History of Creation* (1844), written by the Scottish geologist Robert Chambers, stirred controversy with its radical naturalistic account of life. For Chambers (1844: 305), the first organisms arose through spontaneous generation and evolved over time. Humans all share a common descent, which was "at first in a state of simplicity, if not barbarism," a striking difference from the theological doctrines of the Fall and *imago Dei*, which saw prelapsarian humans as living in a state of original righteousness (as we will see further on).

The *Origin of Species* thus was not the first book to introduce evolution (at the time more commonly referred to as transmutation), but it proved more influential than other transmutationist theories because it formulated a compelling mechanism of how evolutionary change could occur, namely natural selection. Huxley (1863) was the first to apply natural selection to humans: drawing on a rich body of evidence, in particular comparative anatomy and embryology and newly discovered fossils of Neanderthals, he argued that the same evolutionary processes that were at work in other organisms also operated in humans. Darwin (1871) overcame his initial reluctance to write about human evolution and introduced his ideas together with a theory on sexual selection. Like Chambers and Huxley, he proposed a single evolutionary origin for humanity, locating its cradle in Africa. Crucially, Darwin maintained that the continuity between humans and other animals was not merely physical (a position already defended by earlier transmutationists) but also mental, positing precursors to religion and morality in nonhuman animals. As we will see in Section 4, evolutionary explanations of religion posed their own specific challenges to theists.

In the late nineteenth and early twentieth centuries, the scanty hominin fossil record gradually increased. Early paleoanthropologists such as Eugène Dubois and Raymond Dart were keen to find what they considered the "missing link" between humans and primates. Dubois found a *Homo erectus* fossil (at the time dubbed *Pithecantropus erectus*) in Java in 1891 and advocated an Asian origin of humanity. Dart (1925) discovered an *Australopithecus africanus* skull and supported an African genesis of hominins; this find was followed by Broom's discoveries (see below).

At the same time, a metaphysical discussion took place on whether there was any directionality in human evolution and what this entailed for morality and religion. For instance, in a scathing review of *Vestiges*, Sedgwick (1845 [1890]: 84) worried that if transmutationist theory were correct, "the labours of sober induction are in vain; religion is a lie; human law is a mass of folly, and a base injustice; morality is moonshine ... and man and woman are only better beasts!" The Jewish theologian Mordecai Kaplan also was concerned about nihilistic implications of evolutionary theory, worrying that accepting evolution without creation would relieve us from any moral responsibility:

> That God is creator means that the world is not governed by blind necessity. What can exercise a more blighting effect upon all moral endeavor than the notion that there is no meaning or purpose to the world, and that it is soulless in its mechanistic perfection ... We may accept without reservation the Darwinian conception of evolution, so long as we consider the divine impulsion or initiative as the origin of the process. (Kaplan, 1934 [2010]: 98)

One way around the threat of nihilism, especially moral nihilism, was to interpret evolution in progressive terms. For example, Haeckel (1886) designed phylogenetic trees as a way to visualize evolutionary relationships between organisms. His pedigree of humans shows humanity indisputably as the endpoint of the evolutionary process, and other life forms lower down the tree. Contemporary views on human evolution reject this teleology, at least explicitly. Humans are "just another unique species," to use Foley's (1987) memorable phrase. Nevertheless, the tendency to impose a teleological narrative on human evolution remains irresistible. The implicit narrative is one of progress toward "us" with striking parallels to religious origin stories. Misia Landau (1991) has argued that scientific accounts of human origins are structured like a heroic quest narrative, such as the *Odyssey* or *Beowulf*. The plucky hominin ancestor embarks on a difficult quest – becoming human – through the journey of human evolution. Recurring elements of this narrative are terrestrialism (humans descended from the trees to the ground), bipedalism, encephalization, and civilization (social division of labor, sophisticated technology, and art). The quest is riddled with obstacles such as population bottlenecks and climatic challenges (e.g., continental rifting, ice ages), which the hero overcomes. Fortunately, the hero receives gifts to aid him in this quest, such as language, control of fire, and cumulative culture. Anything that does not fit this teleological narrative is subject to intense scrutiny. For example, the small-statured hominin *Homo floresiensis* bucked the trend of encephalization (the gradual increase in hominin brain size). Standing about 1.1 meter tall and with a brain size of ca. 426 cc, these fossils were met with considerable skepticism, especially since the most recent individuals are only 50–60,000 years old (Van den Bergh et al., 2016). Numerous authors (e.g., Martin et al., 2006) proposed that *Homo floresiensis* was not a genuine hominin species but a modern human with a pathology, perhaps microcephaly or Laron's syndrome (see De Cruz & De Smedt, 2013b for discussion). This debate has now been decidedly settled in favor of *H. floresiensis* as a distinct hominin species (Van den Bergh et al., 2016). If it weren't for the reversal of a perceived trend (encephalization increases in the course of human evolution) and its teleological flavor, scrutiny of and skepticism about this species would likely have been less severe.

3.2.2 Contemporary Work on Human Evolution

Contemporary views on human evolution are informed by archaeological and fossil finds, multiple dating methods, and subdisciplines such as paleobotany and paleogenomics. The picture of human evolution is constantly changing, so we focus on a few trends and established findings as well as some new developments. We present the established view that human ancestors evolved in Africa with several sorties out of that continent into Eurasia throughout hominin evolution, and the degree of interbreeding of *Homo sapiens* with populations in Eurasia as a continued topic of research and debate. It is beyond reasonable doubt that humans are closely related to other apes. The split between human and chimpanzee lineages occurred between 4 and 6 million years ago (myr). The earliest hominin fossils in the East African Rift Valley include *Orrorin tugenensis* (6.1–5.7 myr), *Ardipithecus ramidus* (4.4 myr), *Kenyanthropus platyops* (3.5 myr), and *Australopithecus afarensis* (3.9–2.9 myr). These species combined an arboreal lifestyle with bipedalism. Since the late Miocene and early Pliocene (ca. 8–5 myr), the local and global climate was characterized by increased cooling, drying, and instability. This exerted strong selective pressures on a wide variety of mammalian clades, including great apes, hippos, bovids, and suids. Great apes faced increasing competition from the Old World monkeys and adopted a more diversified diet in response. The ancestors of the gorillas specialized in bulky, low-quality food, whereas the common ancestor of hominins and chimpanzees focused on high-calorie diets, including fruit, insects, and occasionally meat (Milton, 1999). Exacerbating aridity meant fruit and leaves became patchier as grasslands emerged. In apes, bipedalism is energy-efficient compared to quadrupedal locomotion (Leonard & Robertson, 1997). This exerted selective pressure on ancestral hominins to adopt a bipedal gait, which was combined with tree climbing to forage for fruits and shoots. The morphology of early hominins reflects this: they were not truly bipedal (that would only happen with *Homo ergaster* at around 1.9 myr), but combined climbing trees with walking. The first stone tools appeared around 3.3 million years ago (Harmand et al., 2015); these were simple cores for extracting marrow by crushing bones and sharp flakes for defleshing bones. Early *Homo* were at the very least confrontational scavengers, chasing predators away from their prey, as is evident from patterns of cutmarks on bones that indicate early access to carcasses that still had a lot of meat on them, and teethmarks on top of the cutmarks left by stone tools, showing scavengers came after hominins (Domínguez-Rodrigo et al., 2014). This same period witnessed two evolutionary trends: a specialist road of increasing robustness and a diet of seeds and nuts (*Paranthropus*), and a generalist high-quality diet reliant on meat (*Homo*). Eventually, the genus *Paranthropus* went extinct.

In the Middle Pleistocene, the genus *Homo* spread across the Old World, with fossils dating to around 2.1 myr in Shangchen (China), 1.8 myr in Dmanisi (Georgia) and Java, and around 1.4–1.2 myr in Spain. Later sorties out of Africa included *Homo heidelbergensis*. Eventually these ancient populations and their descendants were replaced by *Homo sapiens*, although limited gene flow of non-African populations such as *Homo neanderthalensis* in Europe and the Siberian Denisovans contributed to our gene pool (Reich et al., 2010). As it currently stands, the ancient DNA evidence combined with paleoanthropological data supports the following picture: there were multiple exits of hominins out of Africa since at least 2.1 myr. *Homo heidelbergensis* gave rise to *Homo neanderthalensis* and the Denisovans outside Africa, and to our species inside Africa. The earliest *Homo sapiens* fossils are dated to about 315,000 years ago, in Jebel Irhoud, Morocco, and currently no older *Homo sapiens* fossils have been found outside Africa, supporting an out-of-Africa origin for our species (Richter et al., 2017). The earliest exit out of Africa by *Homo sapiens* is about 210,000 years ago (Harvati et al., 2019).

This messy ancestral history, together with fossil and other paleoanthropological data, raises philosophical questions about what a human is and what defines humanity. The term "human" is not a scientific category. Among paleoanthropologists there is no agreement on its scope. Are humans only those hominins who are anatomically modern, or perhaps only those that have art and religion together with technological innovations, as Ian Tattersall (1998) holds? Perhaps we can incorporate other hominin species too, as Robert Foley (1995) does, accepting all hominin species after the human–chimpanzee split as human. These definitional issues matter because scientific theories of human evolution are not purely descriptive – they also make metaphysical claims about what it means to be human. For example, a debate rages on whether Neanderthals were capable of art, body decoration, and language, which in practice often translates into the question of whether they were "like us" or perhaps just some inferior version of us. João Zilhão (2012), an archaeologist who defends complex Neanderthal cognition, argues that the study of Neanderthals is marred by double standards, with much higher standards for evidence placed on *Homo neanderthalensis* than on *H. sapiens*: art objects and complex artifacts at Neanderthal sites are frequently explained away as a result of cultural assimilation or exchange with *H. sapiens*, a move reminiscent of nineteenth-century explanations of the religions of non-Western peoples as products of assimilation with Christianity. Nonetheless, the Iberian Peninsula has yielded increasing evidence for Neanderthal art, including shell beads, marine pigments, and cave paintings with simple geometric lines, dating from well before *H. sapiens* migrated into Western Europe (Hoffmann et al., 2018). In combination with the recognition that both species interbred to some extent (about 4 percent of Eurasian DNA is of Neanderthal origin), this has led to a more positive appraisal of *H. neanderthalensis*.

3.3 Original Sin and the Fall

The doctrine of original sin is a theological rather than a biological position, yet it exerted a profound influence on modern biological thinking. For example, the transmutationist book *Vestiges of Creation* (1844) explicitly pushed back against the idea of a pre-Fall perfection of humanity. We will here examine how paleoanthropology challenges the notions of original sin and the Fall. According to the doctrine of original sin, all humans have an inescapable propensity to sin, which they inherit from their ancestors. In a Christian theological context, sin does not simply mean doing wrong (although in practice it often amounts to that) but rather denotes willful rebellion against God. This transmitted original sin finds its origin in the primal sin committed by the first human couple in the Garden of Eden. In spite of the centrality of original sin in Christian theological anthropology, there is no consensus on how it should be fleshed out doctrinally.

Scriptural references to human origins in the Bible are sparse. The main material is concentrated in the first three chapters of Genesis, which contain two creation narratives and the story of the Fall. The first creation narrative, Genesis 1–2 (dubbed P), is a poetic description of creation in six days, where God speaks things into existence in a systematic manner. Humans are created on the sixth day, after all other animals. The second creation story, older than the first, makes up the bulk of Genesis 2 (dubbed J). It shows God as an improviser who creates the first man ("Adam") before the other animals. God creates animals as companions for Adam, and only later the first woman, named "Eve" by Adam. Genesis 3 presents the Fall narrative. Eve is seduced by a speaking serpent, eats from the forbidden fruit of the Tree of Knowledge of Good and Evil, and convinces Adam to do the same. As a result of their disobedience, death enters the world, humans need to work for their livelihoods, and women experience pain in childbirth and patriarchal subordination. Theologians and biblical scholars do not routinely interpret these creation narratives in a literalist way (for one thing, the creation stories do not agree with each other on the order in which things were created), and the question of the extent to which these narratives were originally meant to be taken literally continues to be a topic of debate (see, e.g., Harris, 2013 for an analysis of the genres to which these narratives belong).

Together, the two creation narratives and the story of the Fall form the basis for two influential doctrines in Christian theology, *imago Dei* and original sin. These are connected: humans bear the image of God, but original sin has marred that image in us, so that one can distinguish two eras in human history, the period before the Fall and the time after. Early Christian writers, including Paul, did not have a substantial theory on original sin. Paul discussed sin (*hamartia*) at length, in particular in his Letter to the Romans. He regarded sin as a corporate and social force that has humans in its thrall. Sin is a slave master (the body of sin) that he juxtaposed with the early church, the body of Christ (Croasmun, 2017). A lot of discussion has centered on the translation of Romans 5:12, where Paul draws an explicit analogy between Christ and Adam: "Therefore, just as sin came into the world through one man, and death came through sin, and so death spread to all because all have sinned." The meaning of this passage is controversial, as it is unclear whether it means that everyone has literally sinned in Adam. Contemporary Pauline scholars tend to deny this interpretation (e.g., Green, 2017), but Augustine, and many historical and contemporary authors following him, have interpreted it as saying that all people have literally sinned in Adam. Pre-Augustinian Church Fathers had quite divergent views on the Fall, which have often been distorted as if they were all leading up to Augustine's influential doctrine of original sin. Augustine developed his position as a response to Pelagianism. Pelagius (ca. 360–418) had an optimistic view on human free choice and sin and believed it was possible for people to attain moral perfection and not sin if they exercised their free will rightly. Augustine worried that this would make grace redundant. His doctrine of original sin is in fact a bundle of five closely connected doctrines (see Couenhoven, 2005 for a detailed treatment). (1) The first human couple was in a state of original righteousness prior to the Fall. They were physically fit and able and also mentally superior to later humans. They were not perfect (since they were still able to sin), but they could choose not to sin if they so wished, something later humans cannot do. (2) But they exercised their free will and disobeyed God by eating from the forbidden fruit. (Augustine believed the Fall was an actual historical event.) This primal sin gave rise to original sin, the sin that is transmitted from Adam to subsequent generations. (3) As a result of original sin, we are weakened and experience death. Not only that, but (4) we also share Adam's guilt for the primal sin. (5) Original sin is transmitted biologically. Augustine thought, in line with the biological understanding of his time, that Adam's semen contained the seeds of successive generations.

Even before theologians became acquainted with evolutionary theory and paleoanthropology, several elements of this doctrine were controversial. Take the idea of original righteousness. If humans were so superior, why did they fall? Positing original righteousness made the Fall deeply puzzling. Either sin is impossible or it was bound to happen because the first human pair already had the intention to disobey God (Duffy, 1988). Augustine favored this latter interpretation, arguing that humans were proud. Before they sinned, they already were guilty of self-aggrandizement and overconfidence (Augustine, 416 [2002], book 11). But other theologians, such as John Hick (1966), find this interpretation implausible, preferring the Irenaean view on original sin.[6] According to the second-century bishop Irenaeus, humans were created immature, like children, and thus were originally innocent rather than righteous. Irenaeus accepted the Fall as a historical event that stalled human moral development. While the first couple was wrong to eat from the forbidden fruit, they were, in Irenaeus's view, victims of the serpent (basically, falling for a scam).

Another part of Augustine's doctrine that attracted theological criticism was original guilt. How can a just God hold descendants accountable for the mistakes of their ancestors? This question remains, regardless of how one sees the propagation of original sin. While Augustine interpreted the transmission of original sin biologically, Calvinists have proposed a federalist interpretation where Adam is the representative of humanity (its federal head), which makes his sin ours. If sin were biologically transmitted, an all-powerful and all-good God could presumably make it the case that it is not transmitted in this way. Similarly, making Adam the federal head of humanity does not solve the problem of justice. It remains equally puzzling why humans today would suffer such grievous consequences of the sin of their federal head if God presumably knew what was going to happen (van den Toren, 2016).

Although the Fall as a historical event came under heavy criticism as a result of evolutionary theory (see below), theologians have objected to its historicity on other grounds. For example, the theologian Friedrich Schleiermacher (1768–1834) argued that the idea of original sin, instigated through a primal sin, is incoherent. The only way to make sense of the primal sin is that the first human beings already had within them a propensity to sin – a perfectly righteous pair would not have sinned. It is therefore more parsimonious to explain each of our sins as a result of this human propensity to sin rather than of a first sin that had a ripple effect through the ages. Moreover, positing a first sin veers

---

[6] Irenaeus's writings on this topic are fragmentary and not as cohesive as Augustine's, which has led to overinterpretation and attribution of views he may not have held. For the relevant passages, see Irenaeus (second century [1884], book 3, chapters 18, 22; book 4, chapter 38) and Irenaeus (second century [1997], chapters 11–16).

dangerously into Manichaean territory, as one would need to explain the human propensity to sin as stemming from a temptation from the devil rather than a misuse of free will. But if one prefers the latter interpretation, there is no need to invoke a first sin to explain why humans sin today (Schleiermacher, 1830 [2016], §72). Although Augustine's doctrine of original sin was challenged on theological grounds long before evolutionary theory came to the fore, evolution brought specific challenges. We now look at Christian authors who considered this doctrine in the light of evolution, and we evaluate it using recent paleoanthropological and paleogenomic evidence.

3.4 Early Evolutionary Responses to the Doctrine of Original Sin

Frederick Tennant (1866–1957), an Anglican theologian with a keen interest in evolutionary theory, examined the implications of evolutionary theory for the Fall and original sin in his *Origin and Propagation of Sin* (1902).[7] To Tennant, human evolution made a historical Fall untenable. Instead, he explained the human propensity to sin as animal tendencies we inherited from our hominin ancestors. Tennant did not dispute that humans were sinful. What he did object to was the idea that a historical act could account for this tendency today, as this would require acceptance of a Lamarckian principle – sin would be an acquired characteristic that was inherited. Instead, he held that sin was a relic of our gradual evolution. Given that humans, in Tennant's view, were natural before they were moral, it becomes inevitable that we sin at some point. Sin is not a falling down, but a failure to rise up morally. For humankind, the demand to rise up morally "makes great demands upon his organic nature, whilst his inherited psychical constitution is making no corresponding or adaptive change, no evolutionary progress" (Tennant, 1902: 102).

One implication of Tennant's view is that God is indirectly responsible for our propensity to sin, as God chose the method of evolution as a mode of creation. This worry was expressed by Tennant's critics, such as E. J. Bricknell (1926), who thought that shifting the responsibility for our tendency to sin to God would let people morally off the hook. Bricknell believed that social models of the transmission of sin could be helpful in this respect, an avenue of research we explore in more detail below. Tennant considered the problem of God being responsible for sin and concluded that while God may be responsible for the possibility of sin, humans are responsible for the actuality of sin (Brannan, 2007). This, however, does not solve the problem. Suppose, to take

---

[7] He wrote two other books on the topic, *The Sources of the Doctrines of the Fall and Original Sin* (1903) and *The Concept of Sin* (1912).

an analogy, that an engineer designed a faulty machine. The machine is complex, capable of performing complicated sequences of actions and making sound decisions, but it cannot but make particular errors in certain situations. A technician causes serious harm while operating this machine out of his own free will. Would not both technician (who committed the harm) and engineer (who designed the machine) appear in a court of law?

One historical way to synthesize evolutionary theory and religious views is to interpret evolution teleologically and in a progressivist manner. For example, Robert Broom (1866–1951) made several palaeoanthropological discoveries, including a near-complete *Australopithecus africanus* skeleton at Sterkfontein and parts of a skull and teeth of *Paranthropus robustus* at Kromdraai (South Africa), thereby helping to establish the case for the African ancestry of humanity. From his published and unpublished writings it is clear that Broom wrestled with questions on metaphysics and religion (Štrkalj, 2003). He practiced medicine but in his spare time conducted scientific research. Science, and particularly paleoanthropology, was not professionalized to the extent that it is today, and scientists often had to take up other jobs next to their scientific work (such as Dubois, who was a military physician). So in the late nineteenth and early twentieth centuries it was not unusual for a medical doctor to dabble in paleoanthropology and yet make significant discoveries. Broom was interested mammalian evolution, particularly in human evolution. While he accepted evolution, he was skeptical of the predominant explanations, in particular Darwin's evolutionary theory. He thought Darwinism could not explain the origins of variation (which indeed remained unexplained prior to the modern synthesis) and that both theologians and Darwinists were dogmatic. He proposed a theory that would combine elements of both. In his view religions (including non-Christian religions such as Judaism and Islam) were a legitimate source of knowledge. In an unpublished manuscript (estimated to be written in late 1947 or early 1948), Broom pondered his discovery of the Sterkfontein *Australopithecus africanus* fossils, which together with other fossils led him to the definite conclusion that evolution has occurred on earth. Evolution was not haphazard but was guided by some intelligent Spiritual Power whose goal was to create humans. The Sterkfontein fossils reflect this work in progress.

> And what was the use of creating man, it may be asked? Pretty surely not the production of a large brained ape that walks on its hind legs and makes war. . . . there is something more in man than in apes – a soul . . . all Evolution on earth has apparently been for the creation of new spiritual entities or souls which will survive death and be used in some spiritual sphere. (Broom, [unpublished] 2003: 129)

Broom believed that a spiritual, intelligent force was at the basis of evolution, in particular, of macroevolutionary changes. If one or more Spiritual Powers created life on Earth, then the end product must not be purely biological but also spiritual. He thus regarded human souls as the *telos* of evolution and believed that "super man," who may arise only 50,000 years from now, would be the final evolutionary step. While we have not reached this stage yet, Broom speculated that some people are able to apprehend this spiritual reality and mentioned Joan of Arc and Gandhi as examples of people who had reached both moral and spiritual superiority (Štrkalj, 2003: 37–38).

3.5 The Doctrine of Original Sin in the Light of Paleoanthropology

Since Tennant and Broom wrote on this topic, there have been changes in debates on human origins and the challenges these pose to the doctrine of original sin. For one thing, we now have more paleoanthropological data and also more theological work that tries to come to terms with this evidence. It is thus worthwhile to revisit the doctrine of original sin. We focus on the historicity of Adam and Eve, the historicity of the Fall, the cognitive consequences of the Fall, the transmission of original sin, and original guilt.

3.5.1 The Historicity of Adam and Eve

The literalist reading of Genesis 1–2 is that all humans are descended from a historical ancestral couple, a reading that many theologians, especially in the Reformed tradition, prefer even if they are nonliteralist in their interpretation of other aspects of Genesis 1–2, such as the chronology of creation (e.g., Jaeger, 2017). Population geneticists look at variability of genomes of present-day humans and extrapolate ancestral population size from this. Estimates of effective population sizes of the ancestral *Homo sapiens* population range from 700 (Zhivotovsky et al., 2003) to about 7,000 individuals (Tenesa et al., 2007). Not a single genetic study concludes that there was just one couple ancestral to all humanity. From genomic evidence we must therefore conclude that there was no historical Adam and Eve.

One way to salvage the historical couple is to posit just one pair as representatives to humanity – this pair was part of a larger ancestral population, an adaptation of the Calvinist federalist model (e.g., Jaeger, 2017). While this position is compatible with the genetic evidence, it still leaves objections about God's justice unanswered. Why did God, presumably knowing what would happen, take this particular couple as representatives of all humanity? Moreover, the position weakens several biblical claims about Adam and Eve being the ancestors of all humans or that their transgression would lie at the basis of biological death or the start of agriculture. For one thing, organisms have been dying since the beginning of life at least 3.5 billion years ago. For another, our species is at least 315,000 years old (Richter et al., 2017) and the oldest forms of agriculture emerged only about 12,000 years ago.

3.5.2 The Fall As a Historical Event

In the Augustinian doctrine of original sin, the Fall is a historical event. Augustine did not read the entirety of Genesis in a literalist fashion, although he did prioritize the literalist interpretation. He favored a figurative interpretation if the literal meaning was absurd (e.g., Adam fashioned from mud), but if a literal interpretation seemed plausible and coherent, it was to be preferred (Augustine, 416 [2002]). But should contemporary authors still accept the Fall as a historical event? James Smith (2017), a philosopher of religion, thinks that a lot hangs on this. The biblical narrative is a plot to which theologians can add "faithful extensions." Some of these extensions preserve the plot, others do not (and thus are less desirable, in Smith's view). To preserve the plot, it is not essential that there is just one Fall by one single couple, but the Fall as an event in time is required for the necessity of grace. Without an original righteous state, we cannot make sense of grace. In the light of this, Smith (2017) proposes an ancestral hominin population that developed cultural transmission, consciousness, and moral capacities, which God elected as his covenant people. They bore the image of God. Unfortunately, they fell, which marked an ontological shift in the human condition. *Pace* Smith, the paleoanthropological record does not show any evidence for a historical Fall or for a period of original righteousness. While violence was less common among ancestral hominins before agriculture and the development of large-scale societies (McCall & Shields, 2008), there are several instances of prehistoric violence. For example, fossils of hominins that lived at Gran Dolina (Spain) 800,000 years ago show butchery marks, an indication of cannibalism. The main victims were children, probably from other hominin groups, witnesses of intergroup violence that still happens among chimpanzees today (Saladie et al., 2012). In nearby Sima de los Huesos 500,000 years ago, a hominin was killed by repeated blows to the head with the same stone tool. Each of these blows was severe enough to be fatal and their repetition indicates an intention to kill. The fact that the lesions do not show any signs of healing suggests the perpetrator was successful (Sala et al., 2015).

The Irenaean view has potentially more leverage, as it proposes original innocence rather than original righteousness. As we have seen, in Irenaeus's view Adam and Eve were like immature children who made a mistake that God sets right through the Incarnation. The Incarnation recapitulates Adam, with

Jesus setting things right where Adam went wrong. Irenaeus did not neglect women in his account, as Eve is recapitulated in Jesus' mother Mary. The Irenaean account still posits a historical Fall, but in this picture, the Fall and the ensuing redemption is not a plan B, but what God intended from the outset.

In the Augustinian picture sin has allegedly large metaphysical consequences, not just for humans but for creation as a whole. We here focus on one domain, the noetic effects of sin. According to Reformed epistemologists (e.g., Plantinga, 2000), sin can explain certain undesirable features of human cognition, collectively referred to as noetic effects of sin. John Calvin (1559 [1960]) proposed that all human beings are endowed with an innate sense of the divine, or *sensus divinitatis*. But due to the noetic effects of sin, the *sensus divinitatis* doesn't always function well, which explains why some people are unable or unwilling to believe in God. While he does not take a stand on the reason why we are in a state of original sin, Plantinga (2000, 207) regards sin as both a cognitive and an affective disorder. One problem with this account is that it conflicts with evolutionary explanations of cognition. Previously we argued that evolutionary approaches to religion provide competing scientific models of why humans have unbelief or myriad incompatible religious beliefs (De Cruz & De Smedt, 2013a). None of these models have a Fall as part of their explanation. Instead, unbelief is the result of a constellation of sociological and cognitive factors. For example, existential security (from untimely death, disease, and poverty) is a factor that decreases religious belief, whereas existential insecurity (loss of personal control, social isolation, and awareness of death) increases it. This explains why countries with low income inequality and good social services tend to have low religiosity (Norenzayan & Gervais, 2013). Beliefs that Christians would regard as incorrect, such as ancestor worship and belief in local place spirits, are predicted by evolutionary and cognitive theories of religion (see next section).

If there were a period of original righteousness followed by sin and its noetic effects, we would expect to see traces of this in the archaeological record. While the prehistoric record shows regional and temporal variations in stone tool technology, explanations for such variations are demographic (smaller or less connected populations are less able to maintain high cultural complexity) rather than religious or moral. What the archaeological record does not show is a period of mental and moral superiority, followed by a permanent decline, as posited by the Fall narrative.

3.5.3 Transmission of Original Sin

Augustine's transmission of original sin is based on seed principles, an ancient understanding of how reproduction works, embedded in his metaphysics of creation. God created the world in two stages: some elements were fully formed (e.g., planets), whereas others (living things) lay dormant within creation as seeds, which would only later spring to fruition. Human beings are latently present as seeds within their ancestors' sperm. If we had literally been present in our ancestors as seeds, sin could potentially be transmitted biologically, though exactly how is still to be fleshed out (Lamoureux, 2015). While Tennant objected to this because it would imply a Lamarckian inheritance of acquired characteristics, he nonetheless believed that sin was a biological propensity, transmitted biologically from nonhuman ancestors. Patricia Williams (2001) provides an updated account of this biological picture. She argues that there was no primal sin but that original sin arises as a result of our conflicted, evolved nature: we are self-interested and selfish but also cooperative and caring.

An alternative theological tradition regards the transmission of original sin mainly in social terms. Schleiermacher outlined a social model of the transmission of sin to solve the following paradox: our tendency to sin is not something that originates in us, but yet somehow we are responsible for our sins. He wanted to avoid some of the difficulties of the Augustinian account, in particular the Fall as a historical event (which he rejected on theological grounds) and original guilt. According to Schleiermacher, knowing the good is prior to sin. We can sin only if we know what is right. To capture this idea of the right he posits God-consciousness, a theological notion that has three distinct elements: self-consciousness, consciousness of how we are socially situated ("species-consciousness"), and consciousness about our relationship to God, which he regards as absolute dependence. Sin is a deliberate resistance to this God-consciousness. Unlike Tennant and Williams, Schleiermacher did not accept that our biological inclinations constitute sinfulness per se. That only arises once people develop God-consciousness. However, he proposed that these biological inclinations constitute the initiator of sin:

> If God-consciousness has not yet developed, there is also not yet any resistance to it; rather, what is present is only a self-focused activity of flesh. . . . in the future this self-focused activity of flesh will indeed become resistance to spirit [i.e., sin], but beforehand it cannot actually be observed as sin, but only as seed of sin at best. (Schleiermacher, 1830 [2016], §67: 405)

How then do people turn away from God? Schleiermacher provided a social model to explain how the transmission of original sin might work: we need more than an appeal to individual free will and susceptibility to sin, as "sinfulness is of a thoroughly collective nature" (§71: 428). We can understand sin as part of the "sphere of life in which the individual belongs most closely"

(§71: 429) and that it is also constituted through shared feelings of "families, relatives, genealogy, peoples," the in-groups to which we belong. Sin is transmitted not only horizontally but also through the generations: "What appears from birth as the susceptibility to sin of a generation is conditioned by the susceptibility to sin of earlier generations and itself conditions the susceptibility to sin of generations yet to come" (§71: 429). Although Schleiermacher traced the origins of sin in our biological nature, he outlined a sophisticated account where sin can be further amplified and transmitted through culture. Throughout *Christian Faith* there are hints of a deep time perspective on religious belief and sinfulness, and on the origins of life and even celestial bodies. Obviously, Schleiermacher was not aware of Darwinian evolutionary theory, but there is good textual and circumstantial evidence to show that he was aware of pre-Darwinian transmutationist theories (Pedersen, 2017: 35–39). For instance, while discussing the doctrine of creation he explicitly stated that "we know fairly certainly of our earth that there have been species on it which are no longer extant and that the present species have not always existed" (Schleiermacher, 1830 [2016], §46: 175). Only transmutationists thought that species could go extinct or arise through natural processes. Moreover, Schleiermacher also endorsed the more controversial idea that stars, planets, and indeed life itself emerged through purely naturalistic processes, not through special divine action, and that simpler life forms gave rise to more complex ones.

His positions about the emergence of God-consciousness in human ancestors, the propensity to sin as part of our biological nature, as well as the social transmission of sin over the generations fit well with contemporary insights about gene–culture co-evolution. Culturally transmitted technologies and ways of life enable humans to colonize hostile environments such as icy deserts and tropical rainforests. We also use culturally transmitted norms to regulate our behavior toward others. Young children are already aware of this normativity and spontaneously protest or criticize an agent who violates the rules (in psychological experiments such agents are usually embodied by puppets). They also understand that norms are context-sensitive (Schmidt et al., 2016). Increasingly, scholars who study human evolution have become aware of the importance of cultural transmission in shaping our behavior, which cannot be explained by genes alone (e.g., Richerson & Boyd, 2005). To explain moral norms and moral violations in humans, we thus need to understand both the biological propensities that can give rise to this (the biological seeds of sin, in Schleiermacherian terms) and the culturally transmitted ways in which these can be expressed or strengthened. There are two ways in which cultural transmission can help us flesh out a model for the social transmission of sin. Culturally transmitted norms and behaviors can be negative (in Schleiermacher's conceptualization, turn people away from God-consciousness), or people can fail to be God-conscious by adhering to harmful beliefs and behaviors because they believe the majority condones them. Social conformity can make the unacceptable acceptable; as Walter Rauschenbusch (1917: 62) put it, "Sin in the individual is shame-faced and cowardly except where society backs and protects it." Conformist and prestige biases can influence the acceptability and prevalence of social attitudes and actions. For example, bullying at schools decreases following interventions by popular children who can operate as antibullying ambassadors – if bullying is no longer cool, people stop doing it (Paluck et al., 2016).

Synthesizing the biological and social models of the transmission of sin, one could outline a gene–culture coevolutionary process. Dual-inheritance theories propose that cultural behaviors can have an effect on the genome by exerting selective pressures. A classic example is milk tolerance. Human infants can digest milk thanks to an enzyme, lactase, which helps to break up milk sugar, lactose. Ancestral humans lost this ability at age four or five. However, some human populations in Europe, the Middle East, Northern Africa, and Northern India have a point mutation in the LCT gene that helps them retain the ability to produce lactase and thus digest milk. Archaeological evidence of cheese strainers appears in the sixth millennium BCE (Salque et al., 2013). Cheese reduces but does not eliminate lactose levels. Ancient DNA of humans of this period shows the point mutation on the LCT gene was not present. However, persistent dietary intake of milk products favored this point mutation in their descendants (O'Brien & Bentley, 2015).

Could a similar gene–culture coevolutionary process be at work in the evolution of morality and sin? Richerson and Henrich (2012) argue that pride, shame, and guilt are social emotions that help us to live harmoniously in groups. Such emotions motivate us to follow norms and dissuade us from violating them, and help us to optimize social behavior. Richerson and Henrich (2012) posit that the evolution of these social emotions allowed us to navigate increasingly complex social environments. Such emotions can give rise to sinful tendencies. For example, pride is an emotion that helps regulate one's sense of social status and that can motivate people to do things that benefit themselves and their group. Joey Cheng et al. (2010) draw a distinction between authentic pride and hubristic pride. Authentic pride is a pride felt in one's own or one's group's accomplishments. It increases generosity and motivates one to do things that increase one's sense of pride. By contrast, hubristic pride is pride about an enduring trait one attributes to oneself or one's group. This latter form of pride is associated with narcissism, arrogance, and a host of other negative emotions, leading to behaviors such as aggression, intimidation, and manipulation. Cheng and coauthors propose that these two forms of pride have evolved to help us navigate social relationships: authentic pride encourages us to do things that enhance prestige, for instance, an excellent chef feels pride in her work and this motivates her to stretch her abilities further, gaining more and more prestige from cooking. By contrast, hubristic pride can lead people to demand deference from others or to despise outgroup members. From this it is clear that sin and morality have both cultural and genetic components.

3.5.4 Original Guilt

Looking at original sin through the lens of human evolution provides the following picture: humans have freedom, but they will nevertheless sin. Would this also make them guilty? Is it reasonable to attribute guilt in a gene–culture coevolutionary model, in other words, can humans be guilty of cognitive proclivities or cultural practices they inherit? There are many situations where people are deemed culpable for beliefs and practices they acquire through cultural transmission. For example, racism and sexism tend to be culturally transmitted, and yet racist and sexist individuals are often held accountable for their beliefs and their actions. There seems to be an evolved basis for favoring people who look similar to oneself, but because humans have free will they are not inevitable racists, which is why they can be held responsible.

Some philosophers have considered the question of blame in relation to implicit attitudes. Neil Levy (2015) recommends holding people responsible only for their explicit attitudes because implicit attitudes are transient and not well-developed, and thus do not track our folk notions of blame and responsibility. We should hesitate before blaming or exonerating someone for their implicit attitudes, and we should not necessarily feel guilty or ashamed about them. By contrast, Jules Holroyd (2012) holds that people can be held accountable for both their explicit and implicit attitudes because they can indirectly exert control over their implicit attitudes. For instance, we can resist reading the kinds of media that foster xenophobia.

Theologians who defend social models of the transmission of sin also argue that people are to some extent blameworthy for the socially transmitted sins they help perpetuate. According to Stephen Duffy (1988: 617), humans can exercise free will but also find themselves in a tragic existential condition where they are part of sinful institutions, which "weave[s] a history which constitutes humanity in its network of interdependence as deaf to the appeal of the good. To be in the world is to be willy-nilly complicit in a sinful condition." Given that we are part of societal structures that systematically perpetuate sin and that we to some extent inevitably partake in (e.g., buying products made in factories with low wages and bad working conditions), we are to some extent guilty of our participation, even though these structures do not originate in us. Schleiermacher (1830 [2016], §71–72) agrees that we are guilty insofar as we perpetuate systemic injustices, but not for inheriting them.

3.6 Conclusion

In this section, we have looked at scientific and theological accounts of human origins, focusing on the Christian doctrine of original sin. These accounts are at odds because of empirical tensions between them and also because they offer conflicting accounts of humanity's place in nature. Contemporary defenders of the doctrine of original sin make empirical claims, but even sophisticated defenses of the Augustinian version are hard to hold up in the light of human evolution. Social accounts of the transmission of sin, such as Schleiermacher's, can be more easily integrated with current scientific insights into the evolution of human cognition and culture.

4 Evolutionary Origins of Religion

4.1 Introduction

There is an enduring suspicion that knowing about the origins of our religious beliefs and practices poses a formidable challenge to them. This nagging distrust underlies seventeenth- and eighteenth-century natural histories of religion and thus predates evolutionary explanations of religion. At the tail end of these origin stories is Hume's (1757 [2007]) *Natural History of Religion*. Hume's argument, following earlier authors such as de Fontenelle (1728), is that religion evolved from polytheism to monotheism. Polytheism emerged when people anthropomorphized their environment. They did this to exert control over a causally opaque world, where they had little influence over natural misfortunes such as draughts or epidemics. Gradually, one of these gods became more widely worshipped than the others because people projected their own political structure, monarchy, on the pantheon. This led to henotheism and eventually monotheism. Natural histories of religion continued with later authors such as Sigmund Freud (1927), who saw religion as an illusion, a projection of the need for a parental figure onto an imagined superpowerful deity.

Contemporary evolutionary explanations of religion are quite different from these earlier accounts. Recent decades witnessed a renaissance of evolutionary, big-picture explanations of religion after the cultural particularism that prevailed in much of the social sciences during the twentieth century. The cognitive science of religion (CSR) is the interdisciplinary study of the cognitive basis of religion. It aims to explain religious beliefs and practices as products of every-day mundane reasoning processes and not as some special domain of human cognition that requires religion-specific explanations. Within this new frame-work it is worth revisiting the question that natural historians of religion pondered: do the evolutionary origins of religious beliefs and practices cast doubt on those beliefs and practices?

We first provide an overview of CSR explanations for religious beliefs and practices and next outline an archaeologically informed account of religion. Archaeology can help generate fruitful hypotheses about the evolutionary origins of religion and, in conjunction with the cognitive sciences, can provide a rich picture of what religious beliefs and practices in ancestral human popula-tions would have looked like. Such pictures require interpretations of the archaeological findings and are subject to potential revisions following new empirical discoveries but nevertheless can further our understanding of how religious beliefs and practices evolved. Having outlined our archaeologically informed account of the evolution of religion, we then briefly examine three general strategies that challenge religion based on evolutionary considerations: appeal to (lack of) sensitivity, (lack of) safety, and sinister genealogy. We show that sinister genealogy accounts are the most doubt-raising for religious beliefs. Evolutionary explanations of religions do challenge some religious views, in particular, that religious diversity would be a result of sin or the Fall or that God would require humans to hold monotheistic beliefs.

4.2 What Phenomena Does CSR Aim to Explain?

Many scholars seem to have a tacit understanding of what religion means and which beliefs and practices count as religious; for example, Wicca is commonly seen as a religion whereas Marxism is not. In spite of this tacit understanding, it is difficult to define religion in a principled way. The Wittgensteinian concept of a family resemblance might be useful: religions have some recurring elements such as belief in supernatural beings or rituals that bring communities together (McDermott, 1970). These recurring elements can be analyzed in an "incre-mental, piecemeal fashion" without worrying about whether they fit into some broader category called religion (Barrett, 2007: 768).

While this piecemeal approach predominates, some CSR scholars have sought to explain religion by reference to a central feature. For example, Pascal Boyer (2002) conceptualizes religion as belief in agents with counter-intuitive properties that violate our intuitive ontologies. Intuitive ontologies are bundles of expectations of how the world works: zombies violate our expecta-tion that the dead are devoid of agency, sacred trees violate our expectation that plants do not understand or care about human behavior, and ghosts violate our expectation that people are not able to walk through walls. Justin Barrett (2000) refined this to minimally counterintuitive (MCI) concepts because folk religious beliefs violate only a limited number of intuitive expectations. To Boyer, supernatural agents are particularly relevant: not only are they arresting because they violate our intuitive ontologies, they also have access to socially relevant information about us. We continuously monitor what others know about us, what they think about us, and what their and our reputation is, so if, for example, one believes the ancestors are closely watching us, these supernatural beings become persons of interest to us. Such supernatural beings are more memorable, hence they have an advantage in cultural transmission, because all things being equal, memorable information has a higher chance of being transmitted. Nevertheless, many religions also include impersonal, teleological forces such as karma or Dao.

Within philosophy of religion, there is a tendency to overintellectualize religion, and this includes the literature on evolutionary debunking arguments as applied to religion.[8] As Sosis and Kiper (2014: 257) note, "if religion is indeed a complex adaptive system that consists of recurring and interacting elements, then the veracity of or warrant for religious beliefs is not challenged by the evolutionary science of religion." This does not mean that religion is immune to evolutionary challenges but that the current philosophical literature does not look sufficiently at religions as practices as well as collections of beliefs. Given the importance of ritual in religious traditions, some authors (e.g., Graham and Haidt, 2010) argue that a proper evolutionary study of religion should focus on ritual and other religious practices rather than on beliefs. But doing so at the expense of analyzing religious beliefs is also lopsided. Religions across cultures come with beliefs in supernatural beings, including monotheistic and polytheistic moralizing gods, place spirits that are worried about ritual but less about moral transgressions, and morally concerned ancestors (e.g., Purzycki et al., 2016). We will therefore adopt a balanced approach that was first proposed by Émile Durkheim (1915): in order to understand the evolutionary roots of religion, we need to look at both religious beliefs and practices.

4.3 Three Assumptions of CSR

CSR is a methodologically and conceptually diverse field that studies evolved cognitive capacities that underlie religious beliefs and practices. It comprises, among others, developmental psychology, sociology, anthropology,

---

[8] By "debunking" philosophers mean to dismiss a view as false or unjustified.

evolutionary psychology, archaeology, and philosophy. In spite of its diversity, three assumptions unify the field.

First, unlike many other approaches to religion, CSR assumes that religion results from ordinary cognitive processes. This assumption is sometimes termed the naturalness of religion thesis. "Natural" is a polysemic concept, but the way CSR authors understand it is that religion arises easily, early in development, and without much cultural input (e.g., McCauley, 2011). That is not the same as saying that religion would arise without any cultural input, as the vast majority of religious beliefs people hold are culturally transmitted and are the result of sophisticated processes of cultural evolution. Rather, it means that religious beliefs can arise without extensive teaching and that they are easy to come up with and transmit culturally. For example, young children spontaneously form the belief that people's mental states continue after they die, and they even think that they themselves existed before they were conceived (Emmons & Kelemen, 2014) – prelife beliefs exist in a number of religions such as Mormonism and reincarnationist religions, but they are by no means universal.

Second, CSR authors assume that in order to explain religion, and culture more broadly, one needs to have an explanation rooted in the cognitive sciences. Cultural transmission and cognitive dispositions are both important to explain why people have the religious beliefs they have. Solely appealing to culture is not enough to explain features of religious beliefs and practices across cultures. Religious beliefs reflect enduring propensities of human cognition. For example, the fact that people believe in an afterlife may be the result of several cognitive dispositions, including the inability to imagine oneself as no longer existing and the ability to think about the mental states of others (De Cruz & De Smedt, 2017). Someone's physical death is no obstacle to continue to attribute mental states to them; we do not need the physical presence of a person to continue to think about what they would say or do, so once that person is separated from us by death, it is straightforward to imagine their continued physical and mental states. However, there are significant cross-cultural variations in how these beliefs are fleshed out within and across religions. For example, Christians in the United States think that psychological properties of a deceased person are more likely to continue after death (e.g., the deceased would still miss her children), whereas Christians in Vanuatu (Melanesia) believe biological properties are more likely to survive than psychological ones (e.g., the deceased would still be able to hear, see, and her heart would still beat) (Watson-Jones et al., 2017).

Third, the cognitive mechanisms involved in religion have an evolutionary origin. At the root of all religions lies an evolved cognitive architecture with dedicated cognitive capacities that play a role in many other everyday activities. Two broad approaches explain religion as a product of evolution: by-product explanations and adaptationist accounts. Stewart Guthrie's anthropomorphism hypothesis is an early by-product account. According to Guthrie (1993), animism is a pervasive feature of perception, not only in humans, but also in other animals. Because perception is interpretive, animals make a bet: they interpret ambiguous signs as having agency because the potential benefits of doing so outweigh the costs of not doing so. Better to mistake a boulder for a bear than a bear for a boulder. In humans, animism expresses itself in the proneness to anthropomorphize the environment, for example, discerning spirits in local features of the landscape such as trees and rivers. As we will see, animism and anthropomorphism form the basis of animistic ontology, which manifested itself in the religions of Upper Paleolithic communities.

Adaptationist explanations hypothesize that religious beliefs and practices provide fitness benefits. Take the *big gods hypothesis*: Ara Norenzayan (2013) aims to explain why humans cooperate in large groups, going beyond reciprocal altruism and altruistic punishment. He argues that such increases in group size were possible thanks to belief in powerful, morally concerned deities (high gods, which Norenzayan terms "big gods"), citing evidence that shows a correlation between belief in high gods and group size (e.g., Roes & Raymond, 2003). While this hypothesis has significant empirical support, it has not controlled for common cultural ancestry. Many high gods probably have a common cultural ancestor; for example, the Abrahamic God currently worshipped in Islam, Christianity, and Judaism spread through a combination of proselytizing, missionary work, conquest, and colonization. When this is taken into consideration, the correlation between group size and high god beliefs is significantly weakened. An alternative hypothesis, coined *broad supernatural punishment*, holds that a variety of supernatural agents, including high gods, place spirits, ancestors, and very constrained supernatural beings such as the Chinese kitchen god, can instill cooperation as long as they have the ability to punish ritual and/or moral transgressions. For example, Watts et al. (2015) looked at a broad range of supernatural agents, including ancestral spirits, in Austronesian (Oceanic) cultures. They found that such agents — and not only supreme creator gods — are able to punish, and thus instill cooperation. Similarly, Purzycki et al. (2016) investigated a cross-cultural sample of belief in supernatural agents, including garden spirits (Vanuatuan horticulturalists), ancestor spirits (Yasawa, Fiji), and spirit masters (Tyva, Siberia). Participants played a game in which they could allocate money either to themselves or to a distant or close person with the same religion. People were more generous to distant coreligionists than to close neighbors of a different religion if the supernatural beings they believed in were more knowledgeable and more capable of punishing moral transgressions. This supports the broad supernatural punishment hypothesis over the big gods hypothesis. Increasingly, the distinction between by-product and adaptationist explanations is blurred; many contemporary CSR theories incorporate elements of both approaches. For example, to Atran and Henrich (2010), religion is an adaptive system that arose through cultural group selection and piggybacks on cognitive predilections such as the belief in MCI agents.

4.4 The Genealogy of Religion: An Archaeologically Informed Approach

We here outline an archaeologically informed genealogy of religious beliefs and practices. It is useful to anchor genealogical accounts of religion in the archaeological record because this guards against just-so storytelling; of course, our picture is subject to revision as more archaeological evidence becomes available. We focus on rituals, belief in an afterlife, and belief in MCI agents, as these elements of religion have left archaeological traces. Archaeology offers mere glimpses of behavior in the remote past; while there is a lot of religious behavior that does not leave any material trace, this is the only *longue durée* perspective we have on the evolution of religious behaviors and beliefs.[9]

4.4.1 Rituals, Cheap and Costly Signaling

The earliest material evidence for religious behavior consists of materials that were used in ritual contexts, including pigments and (later) shell beads, presumably for body decoration. Rituals are not necessarily religious; for example, greeting the flag and company dinners do not have religious content. Ritual helps to increase cooperation through evolutionary ancient mechanisms. For example, synchronous actions, such as dancing and singing, help people to feel more socially bonded; people who sing in large choirs feel more connected with each other; dancing to the same music or even exercising together also enhances social bonding and increases pain tolerance (e.g., Reddish et al., 2013).

Adaptationist theories that explain the role of ritual hold that it is a form of signaling. According to costly signaling theory, participants to religious rituals incur costs in terms of time and energy, and forsake opportunities in terms of diet (food taboos) or engagement with outgroup members (e.g., clothes that mark

---

[9] As a quick guide, here are the prehistoric periods we discuss further on: African Middle Stone Age (MSA) ~300,000–25,000 BP (Before Present), European Middle Paleolithic (MP) ~300,000–50,000 BP, Aurignacian ~45,000–29,000 BP, Gravettian ~29,000–22,000 BP, Solutrean ~22,000–18,000 BP, and Magdalenian ~17,000–11,000 BP.

one as, say, Jewish). Religious practices are honest, costly signals of commitment to a cooperative group. The costs discourage free-riding and hypocrisy, where one would join a religion without the beliefs that sustain such practices, and none-theless enjoy the fruits of cooperation. Empirical evidence for the costly signaling theory includes an analysis of historical communes in the United States, which were more likely to survive if they were religious: strict religious communes (with more prohibitions on food, sex, etc.) were more likely to survive than less strict groups (Sosis & Bressler, 2003). Similar dynamics occur in present-day kibbutzim, where religious kibbutzim are more cooperative than secular ones (Sosis & Ruffle, 2007).

By contrast, cheap signaling theories propose that religious rituals are not necessarily costly. Rituals can help people whose interests are already well aligned to coordinate joint actions better. The archaeologist Steven Kuhn (2014) proposes that ritual started out as cheap signaling to help people coordinate between groups, a cheap way to signal cooperative intent. This could become more costly under specific conditions, for instance, when cooperation is more difficult or the benefits bestowed on ritual participants become more significant.

The archaeological record shows evidence for cheap and costly signaling rituals. One can distinguish three broad waves in the evolution of ritual. The oldest evidence consists of colored pigments in the MSA around 300,000 BP. These pigments have bright, radiant colors, including red, purple, and yellow. Stones containing such pigments, such as hematite and ochre, are found in large concentrations in archaeological sites, brought together from distant sources. The oldest finds come from Wonderwerk cave (South Africa), a cavern of about 100 meters deep that contains abundant pigments, often carried for distances of over 30 km, which is beyond the home range of present-day hunter-gatherer bands. The accumulation of pigments in the cave began 300,000 years ago (Watts et al., 2016). Other early evidence for ritual includes large collections of mineral oxides with traces of rubbing and scraping from Twin Rivers (Zambia) at about 266,000 BP. Colors include yellow, brown, pink, red, and blue-black, and specularite crystals that yield a sparkly purple (Barham, 2002). Ethnographic parallels suggest that these pigments were used for ritual body decoration. Utilitarian explanations, such as the use of ochre for glues in hafting tools, cannot be excluded, but why then would early *Homo sapiens* have curated a wide range of colors and in such large quantities? Since these materials needed to be collected but were neither rare nor difficult to apply, the earliest evidence for ritual supports Kuhn's cheap signaling hypothesis. This earliest evidence of ritual coincides with the oldest evidence for intergroup exchange and collabora-tion, for example, in the exchange of obsidian and pigments around 320,000–305,000 BP in Olorgesailie Basin, southern Kenya (Brooks et al., 2018).

Contemporary hunter-gatherers have exchange networks that help them buffer against uncertainty such as scarcity of food and raw materials, and they engage in ritual and gift-exchange with people from nearby groups to maintain access to food and other necessities (Hill et al., 2014). The archaeological evidence suggests that this pattern of uniquely human cooperation between groups started in the MSA.

A second wave of material evidence is the use of perforated seashells as shell beads starting between 135,000 BP and 90,000 BP in the Levant (Skhul and Qafzeh, Israel), 83,000 BP in North Africa (Grotte des Pigeons, Taforalt, and Ifri n'Ammar, Morocco), and around 75,000 BP in South Africa (Blombos Cave). These shells were worn in strings around the body; in some cases they were covered in red ochre (d'Errico et al., 2009; Bar-Yosef Mayer et al., 2009). Ethnographic parallels such as the *xharo* network among the !Kung (Ju/'hoansi) in southern Africa indicate that wearing particular beads signals membership of mutual assistance networks, where members exchange information about technological innovations and the whereabouts of animals and plants in a ritualized context (Wiessner, 2002). Ethnographically, rituals are also used for information exchange, for instance, about medicinal plants among the BaYaka, Congo (Salali et al., 2016).

A third wave of evidence, which shows an evolution toward costly signaling, can be found in Upper Paleolithic (UP) Europe, roughly between 40,000 and 12,000 BP. Beads in UP sites are often made of rare materials that are difficult to work; for example, one of the most common types of beads in Aurignacian sites is the basket-shaped mammoth ivory bead, which takes an experienced experimental archaeologist at least an hour to carve (White, 1993). The fact that clothing decorated with these beads contains hundreds of them is in line with costly signaling theory. Caves such as Altamira and Chauvet were likely used as aggregation sites, where people of different groups interacted (Conkey, 1980). They contain cave art, which presumably was incorporated in ritual contexts (see below). Children, adolescents, and adult men and women frequented these caves, as can be seen in handprints, footmarks, and finger flutings (tracings of fingers along soft limestone walls) (Van Gelder, 2015). Such age and gender profiles indicate community-wide ritual activities.

4.4.2 Burials and Afterlife Beliefs

While other animals also grieve (see Pierce, 2013 for review), humans are unique in their mortuary practices, which include ritualized mourning and treating the bodies of the deceased in culture-specific ways, such as inhumation or cremation. Mortuary practices can serve several purposes, including dealing with grief, commemorating the dead, and severing social ties with the deceased, whose social relationships (property, marriage, kinship) need to be reconfigured, for example, into those of an ancestor. Mortuary practices also sometimes point to belief in an afterlife, especially if accompanied by grave goods that help the deceased in their postmortem life.

The archaeological record shows a dissociation between burial, belief in supernatural beings, and belief in an afterlife. For example, UP Southwestern Germany (ca. 40,000–30,000 BP) has rich evidence of symbolic material culture, including musical instruments and artifacts depicting MCI supernatural agents (see below), but this region shows no evidence for burial until about 12,000 BP (Sala & Conard, 2016). This indicates that religion did not arise as the package we now know – a combination of moral beliefs, mortuary practices, and belief in supernatural beings – but rather emerged in a mosaic, piecemeal fashion.

There is no firm evidence for burials before 130,000 BP, and burial as a customary practice becomes relatively widespread only after 28,000 BP. (We will here focus on Western Eurasia, as the evidence for burials in the rest of the Old World is sporadic.) To illustrate, only 58 burials are generally accepted for the period 130,000 to 50,000 BP; of those, 35 are Neanderthals (Hovers & Belfer-Cohen, 2013). Before 28,000 BP, evidence for burial remains sporadic, with individual graves rather than group burials. Cemeteries with burials of hundreds of individuals emerge only at the end of the last Ice Age (e.g., Lepenski Vir, Serbia and Natufian culture, Levant).

Mortuary practices do not necessarily point toward belief in an afterlife. Burying the dead might have been an expression of compassion or deference for the deceased: both Neanderthals and *Homo sapiens* cared for injured and disabled individuals, and this care might have been extended to dead group members. Interestingly, the archaeological evidence suggests that Neanderthals started burying their dead only after *H. sapiens* developed the practice, which suggests a pattern of cultural diffusion (Hublin, 2000: 171): the oldest *H. sapiens* burials are around 130,000–92,000 BP (Qafzeh and Skhul caves, Israel), while Neanderthal burials in Eurasia are younger than 70,000 BP.

The presence of grave goods makes afterlife beliefs more likely, although it is impossible to rule out alternative explanations such as costly signaling (the burial of useful goods that are therefore lost to the organizer of the burial). However, since early grave gifts are not exceptional (an animal bone, a shell bead), it might be more parsimonious to accept them as an indicator of belief in an afterlife. Belief in an afterlife, as mentioned earlier, results from several cognitive dispositions, including our spontaneous attribution of mental states, plans, and intentions to others. This seems to warrant the inference that humans around 100,000 BP had a theory of mind similar to ours, ascribing continued existence to departed group members and attributing needs to them in their postmortem life. Grave gifts are already present in early graves such as Qafzeh, which contains ochre and perforated shells. Given the distance to the coast (about 50 km away) and the absence of other marine resources (such as fish bones), it is likely that these shells were obtained through exchange with neighboring groups (Bar-Yosef Mayer et al., 2009). Between 28,000 and 24,000 BP grave goods became more common. Examples include body decoration made of shells and teeth, bone tools and spears, and decorated clothing, for example, in Dolní Věstonice and Předmostí (Czech Republic) and Sunghir (Russia). Archaeology is limited in what it can show, in that it only deals with material remains. Mortuary practices in prehistory undoubtedly involved other behaviors, such as chanting, wailing, and dancing, that are invisible in the archaeological record. But there is a clear trend in increasing inhumation and care for the dead.

4.4.3 MCI Agents

As we saw earlier, Guthrie (1993) proposes that humans have a pervasive tendency to anthropomorphize the environment and to see it as having agency – the animistic ontology that is characteristic of many small-scale societies. Animist ontology can be cultivated and enhanced in religious contexts. Within totemic cultures, people perceive some animals as their ancestors, for example, shark ancestors in the Solomon Islands (Melanesia). In animistic religions, such as the Yup'ik (Alaskan Inuit), other animal species are also perceived as persons and are invited to participate in ceremonies – this does not mean that other animals are perceived as on a par with humans, but that the category "person" is extended to include both humans and nonhuman animals. In such cultures, the distinction between humans and other animals is blurred with art objects, such as animal masks depicting shark ancestors or figurines of bear helper spirits. Therianthropes (human–animal hybrids) can be situated in such religious contexts, where the line between humans and other species is deliberately blurred. About 30 figurines, paintings, and engravings depicting therianthropes have been found in UP sites.[10]

Traditionally, archaeologists (e.g., Conard, 2003) identify figurative art from UP Europe as the earliest firm evidence for religious belief. This includes cave paintings, figurines, and decorated tools such as harpoons and atlatls. While figurative art in contemporary small-scale societies often has a religious meaning, we cannot say with certainty whether this was the case in prehistoric Europe. For example, it is unclear whether the so-called Venus figurines (nude, mostly

---

[10] Count by the authors; list available upon request.

anonymized female figures common in the Gravettian and Magdalenian periods) represent religious concepts. There are many interpretations for these figurines, ranging from self-portraits, erotica, purely artistic expressions, and goddesses (Nelson, 1990). Nonetheless, some UP artworks can be plausibly interpreted as religious, thanks to stable features of human cognition and perception that have an influence on art across cultures.

Some artworks reflect MCI properties, usually a mix of human and animal bodies. One of the oldest sculptures is a mammoth ivory figurine with a human body and a cave lion's head, dated to about 39,000–41,000 BP, found in Hohlenstein Stadel cave (Southwestern Germany). This "lion man" (*Löwenmensch*) stands about 31 cm tall (exceptionally large for a mammoth ivory figurine, which typically does not exceed 5 cm) and has a rigid, upright posture and incisions on its left arm, probably indicating scarifications or tattoos (Kind et al., 2014). A second lion man was found in the nearby Hohle Fels cave in a younger layer, rich in stone, antler, and ivory artifacts, dated to 36,000–33,000 BP. It is smaller (only 2.5 cm) and less detailed, but similar in its rigid, angular pose and head shape. In spite of the chronological gap, these stylistic similarities are remarkable and may point to a continuing cultural tradition (Conard, 2003). In CSR terms, the lion men depict MCI beings because they violate our intuitive biology (as they are a hybrid of two unrelated species, *Homo sapiens* and *Panthera spelaea*) and thus become more memorable – their depiction is more likely to be transmitted. Other examples of therianthropes include a woman–bison hybrid (Grotte Chauvet, France, over 30,000 years old) and a schematic human with two horns (Grotta di Fumane, Italy, 35,500 years old).

4.4.4 Shamanism

Next to large terrestrial animals and hand stencils, UP cave paintings are characterized by the frequent presence of entoptic signs, such as cross-hatchings, dots, and lines. Since the nineteenth century, archaeologists inspired by ethnographic parallels of cave and other art in small-scale societies world-wide have interpreted a subset of cave paintings as made in the context of shamanic rituals. In the mid-twentieth century, this interpretation fell into disfavor due to racist notions of inherent superiority, which depreciated the cultural expressions of non-Western, small-scale societies as meaningless and unsophisticated. Since the 1980s, however, a renewed interest in non-Western religious and indigenous knowledge practices such as ethnobotany and eth-noastronomy brought new life into the shamanic interpretation. Using parallels between historical San cave art from southern Africa and rituals in contempor-ary San religions, David Lewis-Williams (2002) has argued that stable features of human cognition give rise to hallucinations when people experience altered states of consciousness. Area V1, the early visual system, generates the perception of zigzags, lines, lattices, tunnels, and dots (Froese et al., 2013). In later stages of altered states of consciousness, a shaman or initiate may also experience more vivid and detailed auditory and visual hallucinations, morphing to complex scenes and self-transformations. Jean Clottes and David Lewis-Williams (1996) argue that some of the UP caves were used for shamanic rituals, particularly because they contain rock art that depicts entoptic signs and therianthropes. However, Jeremy Dronfield (1996) has shown that the depiction of entoptic signs does not always take place in shamanic contexts, is not always done by shamans, and that some figurative motives are not entoptic signs that are generated by V1. The dappled horses of Pech Merle (France, 25,000 BP) illustrate the difficulties in interpreting prehistoric art: some researchers (e.g., Alpert, 1997) interpret the spots on the horses as entoptic signs, though geneticists (e.g., Pruvost et al., 2011) claim they could be realistic portraits of predomestic horses, showing a genotype expressed in a dappled coat pattern. Depictions of shamans appear only in the late Solutrean and Magdalenian. Examples include a horned, dancing figure from Gabillou cave (France, 17,000 BP) and a dancing human playing a musical bow dressed with a bison head and tail from Grotte des Trois Frères (France, 13,000 BP).

Our archaeological picture is corroborated by Peoples et al. (2016), who used phylogenetic trees to analyze the evolution of religious beliefs among 33 hunter-gatherer societies across the globe, looking at patterns of common ancestry and the absence or presence of the following religious traits: animism, belief in an afterlife, shamanism, ancestor worship, and belief in high gods. They found that animism appeared earliest (present in all societies studied), followed by afterlife beliefs, and only then shamanism. Notably, Peoples et al. (2016) found no evidence for high god beliefs among ancestral hunter-gatherers; their religious beings reflected their social structures, which were egalitarian. While their analysis does not refer to the archaeological record, it coheres with our archaeological approach, which shows animism in the Aurignacian, afterlife beliefs (with habitual grave goods) starting 28,000–24,000 BP, and shamanism (depictions of shamans dating from the Solutrean/Magdalenian) only arising later at 18,000–13,000 BP.

4.5 Debunking Religion: Sensitivity, Safety, or Sinister Genealogy

As we have just seen, the archaeological evidence shows that religion did not evolve as a neat package but instead had a mosaic evolution. Traces of religion appear, disappear, and reappear throughout the archaeological record starting around 300,000 BP. What are the normative implications of the evolution of religion for religious beliefs and practices? The question of how the evolutionary origins of our beliefs can undermine those beliefs has recently received ample attention. Evolutionary debunking arguments (EDAs) have the following general structure: they start out with a premise about the influence of evolution on our beliefs in a particular evaluative or factual domain, then employ various epistemic principles to show that this casts doubt on these beliefs and conclude that we are not justified in holding them. We here focus on three epistemic principles that are used in EDAs: sensitivity, safety, and sinister genealogy.

Sensitivity arguments claim that human brains generate religious beliefs regardless of whether these beliefs refer to anything outside the brain. For example, humans are prone to believe in MCI beings because these concepts are memorable and easy to transmit culturally. These facts about how the human mind functions explain why we are susceptible to believing in the existence of MCI beings. Whether they exist in reality doesn't make a difference to the way we form these beliefs – our brains tend to generate and transmit them. Matthew Braddock (2016) reviews content biases identified by CSR, such as the memorability of MCI concepts, mind/body dualism, and intuitive teleology, and argues that because of these biases, people will end up with a variety of unreliable religious beliefs, at least when gauged against theologically correct concepts of the Anselmian omni-God within Abrahamic traditions. Given that the human brain generates beliefs in multiple gods (such as finite gods, gods who don't care about our moral behavior, gods who are morally concerned), we should suspend judgment about the reliability of the mechanisms that give rise to these beliefs. Note that Braddock does not assume that these mechanisms are unreliable, which would be question-begging, as he sets out to establish their unreliability. Rather, he argues that CSR gives us reason to believe people will naturally end up with mistaken god beliefs, at least according to the standards of the Abrahamic religions.

One might wonder whether CSR is doing any extra work here beyond the challenges from religious diversity that historical authors such as Calvin and the Muslim theologian al-Ghazālī (ca. 1058–1111) already faced. Contrary to Calvin, who claimed our *sensus divinitatis* is marred by sin, al-Ghazālī appealed to *fiṭrah*, a natural belief in God and basic moral sense that can give rise to authentic religion (or be perverted into false religions). Enculturation into one's parents' religion can lead to a perversion of authentic religion (Islam, according to al-Ghazālī): "Every infant is born endowed with the *fiṭrah*: then his parents make him Jew or Christian or Magian [Zoroastrian]" (al-Ghazālī, ca. 1100 [2006], 19–20). Moreover, cultural transmission can tweak the content biases

Braddock discusses to fit within any given religious tradition. As we have seen, young children voice the spontaneous belief that they existed before they were born, so-called prelife beliefs (Emmons & Kelemen, 2014). If children grow up in traditions that endorse such beliefs, such as Mormonism, Buddhism, or Hinduism, these spontaneous beliefs will be reinforced and culturally elaborated. For a sensitivity-EDA to work, one would need to demonstrate that both biological and cultural factors that jointly give rise to religious beliefs would be insensitive to religious truths. However, CSR has not paid sufficient attention to cultural evolution or gene–culture coevolution to sustain this claim.

EDAs based on *safety* start from the assumption that our beliefs, in order to count as knowledge, must be safe from error. Given the evolutionary history of our species, we ended up with specific religious beliefs and behaviors that could easily have been different. If our evolution had been similar to that of eusocial insects, we would have had different moral beliefs. As Darwin (1871: 73) put it, "our unmarried females would, like the worker-bees, think it a sacred duty to kill their brothers, and mothers would strive to kill their fertile daughters; and no one would think of interfering." Similarly, our religious beliefs would have been different given a divergent evolutionary makeup, as already remarked by the Greek philosopher Xenophanes (sixth to fifth century BCE [1898]: 67), "if cattle or lions had hands, so as to paint with their hands and produce works of art as men do, they would paint their gods and give them bodies in form like their own – horses like horses, cattle like cattle." Counterfactually, if other animals had minds capable of forming religious beliefs, their gods would be very different from ours: tiger gods would have no compassion, ant gods would punish love or even sympathy for brothers or pregnant sisters, and shark gods would hail voracity. This would give rise to very different religious beliefs from the ones that humans hold. As Claire White (2018: 41) puts it starkly, "if humans did not have the ability to represent agents as possessing mental states like desires and intentions, then the act of praying to a deity to change an outcome (i.e., supplication, or petitioning) would presumably not have appeared in human culture."

A safety-based EDA against religion could go as follows: our religious beliefs are contingent upon evolutionary processes that could easily have led us to very different kinds of religious beliefs or to no religious beliefs at all. Safety from error is a requirement for justified belief; therefore, our religious beliefs are unjustified. However, Tomás Bogardus (2016) has challenged safety as a requirement for knowledge or justified belief, as many of our beliefs seem justified, yet are formed in an unsafe way. Take a child who learns that lettuce is soporific through reading Beatrix Potter's *The Tale of the Flopsy Bunnies* (1909). She knows this even though this belief was formed in an unsafe way (reading a children's book), yet it counts as knowledge. Her belief that lettuce is a soporific is also justified in spite of its lack of safety – lettuce juice indeed has sedative properties, and Potter likely knew this, since this was widely publicized in the nineteenth century (e.g., Duncan, 1813). An evolutionary theist might hold that our beliefs are presumably safe because God has guided the evolutionary process in such a way that creatures with beliefs like ours would appear. Within this framework, it is unsurprising that God would have used a specific evolutionary pathway to furnish our minds with the religious beliefs they hold.

A third form of EDA appeals to *sinister genealogies* in the formation of religious beliefs. Sinister genealogies can be defined as follows:

> *SG* A sinister genealogy is a contributing causal factor to one's belief that p, which, in the absence of defeater-defeaters or defeater-deflectors, removes one's justification for one's belief that p.

This process may not be the only causal determinant, but it needs to be a key contributing factor. A defeater-defeater is a defeater for SG such that SG does not in fact remove one's justification for one's belief that p. A defeater-deflector prevents SG from being a defeater in the first place. Andrew Moon (2017) offers the following example to clarify the difference between them. You see a red device in a factory. However, the foreman informs you that it is red because all widgets are lit with a red lamp (this provides a defeater for your original belief that the device is red). A friend later tells you that the foreman likes to play tricks on visitors, so you now receive a defeater-defeater and can trust your original visual impression (that the widget is red). A defeater-deflector works similarly, except that your friend warned you in advance that the foreman likes to play tricks on visitors. Now the defeater (the foreman's testimony to the red light) is not even a defeater, because it is deflected by the defeater-deflector in advance.

A historical SG is Bernard de Fontenelle's *Origine des fables* (1684 [1824]). While ostensibly aimed at discrediting Greek mythology, enlightened readers understood that de Fontenelle's target was Christianity. He outlined two main reasons for why myths were invented: ignorance about the true causes of events, and our mental attraction to surprising stories, which we have a tendency to further embellish. He hypothesized an early human who wonders where a river might originate from, and who considers jugs to pour water. No human is strong or big enough to pour a water stream that vast, so he imagines that a god – similar to us, but bigger and more powerful – is pouring the river from a gigantic jug. Hence, a myth is born. Any natural events that humans cannot produce, such as storms, lightning, and ocean waves, are imagined as produced by beings more powerful than we are. This form of analogical reasoning is further embellished upon:

> When we tell something surprising, our imagination warms to its object, and by itself enlarges what it is talking about and adds what would be missing to make it completely marvelous, as if it [our imagination] would regret leaving a beautiful thing imperfect. Moreover, we are flattered by the emotions of surprise and admiration that we cause among our listeners, and so it becomes natural to add upon these as it [admiration] gratifies our vanity. (de Fontenelle, 1684 [1824]: 295, our translation)

de Fontenelle argues that the combination of imagination and vanity would even make an honest person invent or embellish all sorts of stories, and that this gives rise to myths. This theory foreshadows later accounts in CSR, including Boyer's theory of religious concepts as MCI. Does de Fontenelle provide an SG for the origin of mythology? This appears to be the case, given that myths originated long ago, when people had less knowledge about the world than they have now – myths are born from ignorance. Moreover, human imagination makes them even more exotic and alluring, and less truth-conserving. Myths are more likely to be culturally transmitted due to their MCI elements, not because they recount something that actually happened. In the absence of defeaters, learning this would remove any justification one had for the belief in myths. The force of SG accounts thus depends on the details of the historical processes involved: rather than high-level considerations such as sensitivity and safety, we should scrutinize the historical processes and examine whether these are conducive to knowledge or justified belief. People invented myths because of a lack of knowledge. Worse, they even inflated these myths with fantastical details. As a result, de Fontenelle's SG removes justification from the belief in myths.

A defeater-defeater would be that while myths in general are not formed in a reliable way, the myths that one would want to defend (e.g., Christian creation and incarnation myths) are true because they were formed through divine revelation. Plantinga (2000) does something along these lines as he dismisses an SG by Freud. Against Freud's claim that religious belief is a form of wishful thinking, Plantinga argues that even if wishful thinking is in general not truth-conducive, it could still be truth-conducive in the special case of belief in God – perhaps God designed us in such a way that we have a desire for him. This design plan provides a defeater-defeater for the Freudian claim. Plantinga also has a defeater-deflector, the noetic effects of sin: because sin has marred our cognitive capacities, we have distorted, incorrect religious beliefs. Indeed, we can expect such incorrect beliefs to crop up. And how far can such defeater-defeaters and defeater-deflectors be generalized? Could we take any SG for religion and then say that God intended it this way? This seems unlikely because assumptions about God's goodness (which are part of classic omnitheism) pose limits on how God could bring religious beliefs about. Moreover, a defeater-deflector like the noetic effects of sin risks becoming a universal acid that also attacks Plantinga-style models of proper function: if the noetic effects of sin are so pervasive, how can we maintain that our cognitive processes generate correct religious beliefs?

Let's consider the archaeologically informed genealogy of religion outlined in this section. Is it sinister? If the adaptationist explanation of religion is correct, then religious beliefs and practices (which emerged through a combination of biological and cultural evolution) are maintained at least in part due to their fitness benefits on a group level. Rituals were and are maintained because they help people to cooperate better, through either cheap or costly signaling. Burials and MCI beliefs appeared later and are by-products of human cognition, including social attachments, perception, and memory. The archaeological evidence, combined with CSR, seems to suggest that religious beliefs and practices became an enduring part of human life. Under some exclusivist conceptions of an Anselmian omni-God, who expects and demands correct religious beliefs, this genealogy is sinister, as the earliest expressions of religion (MSA rituals) were not belief-centered but were concerned with maintaining relationships with neighboring groups. Later expressions of religiosity such as the therianthropes of UP cave and mobiliary art do not represent theologically correct conceptions of the Abrahamic God but a wide variety of religious beings, including lion men and bison people. These early representations are also difficult to combine with the idea that religious diversity and "incorrect" God concepts are a result of the Fall or that people had "correct" religious beliefs prior to the Fall, as Augustine held (see Section 3) – prehistoric hunter-gatherers did not have any high god beliefs.

Under other religious conceptions (even if we confine them to classical theism), our archaeologically informed genealogy is less sinister. For example, some forms of theism, such as many strands of Judaism, put more emphasis on practices than on beliefs. In these traditions, the evolutionary primacy of religious practice over belief is less surprising. Our genealogy is also compatible with models of theism that see human–divine relationships as slowly developing works-in-progress, such as Irenaeus's model. Under such conceptions, we would not expect a homogeneous monotheism to emerge in early humans. Indeed, as we saw, high god beliefs were very likely absent among ancestral hunter-gatherers.

4.6 Concluding Thoughts

In this Element we have looked at the challenge of evolution to religion along several lines, including the contingency and convergence of evolutionary processes, human evolution, and the evolution of religion. We have chosen not to focus on a blanket challenge of evolution, considered as a whole, and religion, considered as a whole. Rather, we have examined specific aspects of evolutionary theory and how these affect particular religious views. Overall, we found that religious traditions (especially Christianity and Judaism) provide a number of resources that counter challenges that evolutionary theory poses, including the apparent lack of teleology in evolutionary processes, and the competing accounts that the sciences and religions paint of human origins and religious beliefs and practices. Tensions remain, however, and some of the solutions that theologians have proposed to explain human wrongdoing and religious diversity, for example, are not tenable in the light of evolution.

In terms of future directions, given that the challenge of evolution to religion in general is not a solvable question because of the diversity and incompatibility of religious traditions, more work needs to be done to examine the implications of evolutionary theory for nontheistic religious views, such as Daoism, Confucianism, and some schools of Buddhism. For example, Philip Ivanhoe (2017) has recently examined conceptions of the interconnectedness of humans and the natural world in neo-Confucian philosophy, but a thorough examination of whether this oneness hypothesis aligns well with evolutionary theory still remains to be conducted. We expect that future research on evolutionary challenges to religious beliefs and practices will involve detailed investigations, and look at a wider range of religious traditions than the ones currently examined by philosophers of religion and other scholars in science and religion. An expansive naturalism, which does not see religious beliefs and naturalism as incompatible, allows religious believers to seriously engage with both the particulars and theological ramifications of their religious views, and the empirical details and broader metaphysical implications of evolution, including the evolution of their religions.

References

al-Ghazālī, A. H. M. (1100 [2006]). *Deliverance from error* (R. J. McCarthy, trans.). Louisville, KY: Fons Vitae.

Alexander, D. (2008). *Creation or evolution: Do we have to choose?* Oxford: Monarch Books.

Alpert, B. O. (1997). The meaning of the dots on the horses of Pech Merle. *Arts*, **2**, 476–90.

Amundson, R. (2000). Against normal function. *Studies in History and Philosophy of Biological and Biomedical Sciences*, **31**, 33–53.

Aquinas, T. (thirteenth century [1998]). *Aquinas: Selected philosophical writings* (T. McDermott, ed.). Oxford: Oxford University Press.

Atran, S., & Henrich, J. (2010). The evolution of religion: How cognitive by-products, adaptive learning heuristics, ritual displays, and group competition generate deep commitments to prosocial religions. *Biological Theory*, **5**, 18–30.

Augustine. (416 [2002]). The literal meaning of Genesis (E. Hill & M. O'Connell, trans.). In J. E. Rotelle (ed.), *On Genesis* (pp. 155–506). New York: New City Press.

Augustine. (fifth century [1972]). *The city of God against the pagans* (W. M. Green, trans.). Cambridge, MA: Harvard University Press.

Aurobindo. (1914–1918 [2005]). *The life divine.* Pondicherry: Sri Aurobindo Ashram Press.

Banerjee, K., & Bloom, P. (2014). Why did this happen to me? Religious believers' and non-believers' teleological reasoning about life events. *Cognition*, **133**, 277–303.

Barbour, I. (2000). *When science meets religion: Enemies, strangers, or partners?* New York: HarperCollins.

Barham, L. S. (2002). Systematic pigment use in the Middle Pleistocene of South-Central Africa. *Current Anthropology*, **43**, 181–90.

Barr, S. M. (2009). The concept of randomness in science and divine providence. In J. Seckbach & R. Gordon (eds.), *Divine action and natural selection. Science, faith and evolution* (pp. 465–78). Singapore: World Scientific.

Barrett, J. L. (2000). Exploring the natural foundations of religion. *Trends in Cognitive Sciences*, **4**, 29–34.

Barrett, J. L. (2007). Cognitive science of religion: What is it and why is it? *Religion Compass*, **1**, 768–86.

Bartholomew, D. J. (2008). *God, chance and purpose*. Cambridge: Cambridge University Press.

Bar-Yosef Mayer, D. E., Vandermeersch, B., & Bar-Yosef, O. (2009). Shells and ochre in Middle Paleolithic Qafzeh Cave, Israel: Indications for modern behavior. *Journal of Human Evolution*, **56**, 307–14.

Beatty, J. (2006). Replaying life's tape. *Journal of Philosophy*, **103**, 336–62.

Behe, M. J. (1996). *Darwin's black box: The biochemical challenge to evolution*. New York: Free Press.

Blumenbach, J. F. (1781). *Über den Bildungstrieb und das Zeugungsgeschäfte*. Göttingen: Johann Christian Dieterich.

Bogardus, T. (2016). Only all naturalists should worry about only one evolutionary debunking argument. *Ethics*, **126**, 636–61.

Boulter, S. (2019). *Why medieval philosophy matters*. London: Bloomsbury Academic.

Boyer, P. (2002). *Religion explained. The evolutionary origins of religious thought*. London: Vintage.

Braddock, M. (2016). Debunking arguments and the cognitive science of religion. *Theology and Science*, **14**, 268–87.

Brannan, D. K. (2007). Darwinism and original sin: Frederick R. Tennant's integration of Darwinian worldviews into Christian thought in the nineteenth century. *Journal for Interdisciplinary Research on Religion and Science*, **1**, 187–217.

Bricknell, E. J. (1926). Sin and the Fall. In E. G. Selwyn (ed.), *Essays Catholic and critical* (pp. 205–24). London: Society for Promoting Christian Knowledge.

Brooks, A. S., Yellen, J. E., Potts, R., et al. (2018). Long-distance stone transport and pigment use in the earliest Middle Stone Age. *Science*, **360**, 90–4.

Broom, R. ([unpublished] 2003). South Africa and Evolution. In G. Štrkalj and B. Sherman (eds.), South Africa and Evolution: An unpublished manuscript by Robert Broom (pp. 125–30). *Annals of the Transvaal Museum*, **40**, 123–30.

Brown, C. M. (2008). The design argument in classical Hindu thought. *Journal of Hindu Studies*, **12**, 103–51.

Brown, C. M. (2012). *Hindu perspectives on evolution. Darwin, Dharma and design*. London: Routledge.

Calvin, J. (1559 [1960]). *Institutes of the Christian religion* (F. L. Battles, trans.). Philadelphia: Westminster Press.

Casler, K., & Kelemen, D. (2008). Developmental continuity in teleo-functional explanation: Reasoning about nature among Romanian Romani adults. *Journal of Cognition and Development*, **9**, 340–62.

Chakraborty, P. (2001). Science, morality, and nationalism: The multifaceted project of Mahendra Lal Sircar. *Studies in History*, **17**, 245–74.

Chambers, R. (1844). *Vestiges of the natural history of creation*. London: John Churchill.

Cheng, J. T., Tracy, J. L., & Henrich, J. (2010). Pride, personality, and the evolutionary foundations of human social status. *Evolution and Human Behavior*, **31**, 334–47.

Cherry, S. (2003). Three twentieth-century Jewish responses to evolutionary theory. *Aleph: Historical Studies in Science and Judaism*, **3**, 247–90.

Clottes, J., & Lewis-Williams, D. (1996). *Les chamanes de la préhistoire*. Paris: Éditions du Seuil.

Conard, N. J. (2003). Palaeolithic ivory sculptures from southwestern Germany and the origins of figurative art. *Nature*, **426**, 830–2.

Conkey, M. W. (1980). The identification of prehistoric hunter-gatherer aggregation sites: The case of Altamira. *Current Anthropology*, **21**, 609–30.

Conway Morris, S. (2003). *Life's solution. Inevitable humans in a lonely universe*. Cambridge: Cambridge University Press.

Couenhoven, J. (2005). St. Augustine's doctrine of original sin. *Augustinian Studies*, **36**, 359–96.

Croasmun, M. (2017). *The emergence of sin: The cosmic tyrant in Romans*. Oxford: Oxford University Press.

Cummins, R. (2002). Neo-teleology. In A. Ariew, R. Cummins, & M. Perlman (eds.), *Functions: New essays in the philosophy of psychology and biology* (pp. 157–72). New York: Oxford University Press.

Dart, R. A. (1925). *Australopithecus africanus*: The man-ape of South Africa. *Nature*, **115**, 195–9.

Darwin, C. (1860). Letter to Asa Gray, 22 May 1860, DCP-LETT-2814. Accessed on September 26, 2018, www.darwinproject.ac.uk/DCP-LETT-2814.

Darwin, C. (1871). *The descent of man, and selection in relation to sex*. London: John Murray.

Dawes, G. W. (2016). *Galileo and the conflict between religion and science*. London: Routledge.

De Cruz, H., & De Smedt, J. (2013a). Reformed and evolutionary epistemology and the noetic effects of sin. *International Journal for Philosophy of Religion*, **74**, 49–66.

De Cruz, H., & De Smedt, J. (2013b). The value of epistemic disagreement in scientific practice. The case of *Homo floresiensis*. *Studies in History and Philosophy of Science Part A*, **44**, 169–77.

De Cruz, H., & De Smedt, J. (2015). *A natural history of natural theology: The cognitive science of theology and philosophy of religion*. Cambridge, MA: Massachussetts Institute of Technology Press.

De Cruz, H., & De Smedt, J. (2017). How psychological dispositions influence the theology of the afterlife. In Y. Nagasawa & B. Matheson (eds.), *The Palgrave handbook of the afterlife* (pp. 435–53). Basingstoke: Palgrave Macmillan.

de Fontenelle, B. L. B. (1728). *Histoire des oracles* (Revised ed.). La Haye: Gosse & Neaulme.

de Fontenelle, B. L. B. (1684 [1824]). De l'origine des fables. In *Oeuvres de Fontenelle* (pp. 294–310). Paris: J. Pinard.

de Maillet, B. (1748). *Telliamed ou entretiens d'un philosophe indien avec un missionnaire françois sur la diminution de la mer, la formation de la terre, l'origine de l'homme, & c.* Amsterdam: Honoré & Fils.

d'Errico, F., Vanhaeren, M., Barton, N., et al. (2009). Additional evidence on the use of personal ornaments in the Middle Paleolithic of North Africa. *Proceedings of the National Academy of Sciences USA*, **106**, 16051–6.

Dobzhansky, T. (1973). Nothing in biology makes sense except in the light of evolution. *American Biology Teacher*, **35**, 125–9.

Domínguez-Rodrigo, M., Bunn, H. T., & Yravedra, J. (2014). A critical re-evaluation of bone surface modification models for inferring fossil hominin and carnivore interactions through a multivariate approach: Application to the FLK Zinj archaeofaunal assemblage (Olduvai Gorge, Tanzania). *Quaternary International*, **322**, 32–43.

Dronfield, J. (1996). The vision thing: Diagnosis of endogenous derivation in abstract arts. *Current Anthropology*, **37**, 373–91.

Duffy, S. J. (1988). Our hearts of darkness: Original sin revisited. *Theological Studies*, **49**, 597–622.

Duncan, A. (1813). Observations on the preparations of soporific medicines from common garden lettuce (*Lactuca sativa*). *New England Journal of Medicine and Surgery*, **2**, 78–84.

Durkheim, E. (1915). *The elementary forms of the religious life: A study in religious sociology* (J. W. Swain, trans.). London: Allen & Unwin.

Ellis, F. (2014). *God, value, and nature*. Oxford: Oxford University Press.

Emmons, N. A., & Kelemen, D. (2014). The development of children's prelife reasoning: Evidence from two cultures. *Child Development*, **85**, 1617–33.

Evans-Pritchard, E. E. (1937 [1965]). *Witchcraft, oracles and magic among the Azande*. Oxford: Clarendon Press.

Falcon, A. (2015). Aristotle on causality. Stanford Encyclopedia of Philosophy, Accessed on September 15, 2018, from https://plato.stanford.edu/entries/aristotle-causality/.

Finlayson, C. (2009). *The humans who went extinct: Why Neanderthals died out and we survived*. Oxford: Oxford University Press.

Foley, R. (1987). *Another unique species: Patterns in human evolutionary ecology*. Harlow: Longman.

Foley, R. (1995). *Humans before humanity. An evolutionary perspective*. Oxford: Blackwell.

Freud, S. (1927). *Die Zukunft einer Illusion*. Leipzig, Wien & Zürich: Internationaler Psychoanalytischer Verlag.

Froese, T., Woodward, A., & Ikegami, T. (2013). Turing instabilities in biology, culture, and consciousness? On the enactive origins of symbolic material culture. *Adaptive Behavior*, **21**, 199–214.

Garwood, C. (2008). *Flat Earth. The history of an infamous idea*. London: Pan Macmillan.

Gellman, J. Y. (2009). God and chance. In J. Seckbach & R. Gordon (eds.), *Divine action and natural selection. Science, faith and evolution* (pp. 449–62). Singapore: World Scientific.

Godfrey-Smith, P. (2017). The subject as cause and effect of evolution. *Interface Focus*, **7**, 20170022.

Gould, S. J. (1989). *Wonderful life. The Burgess Shale and the nature of history*. London: Penguin.

Gould, S. J. (2001). Nonoverlapping magisteria. In R. Pennock (ed.), *Intelligent design creationism and its critics. Philosophical, theological, and scientific perspectives* (pp. 737–49). Cambridge, MA: Massachussetts Institute of Technology Press.

Graham, J., & Haidt, J. (2010). Beyond beliefs: Religions bind individuals into moral communities. *Personality and Social Psychology Review*, **14**, 140–50.

Green, J. B. (2017). Adam, what have you done? New Testament voices on the origins of sin. In W. Cavanaugh & J. Smith (eds.), *Evolution and the Fall* (pp. 98–116). Grand Rapids, MI: Eerdmans.

Greif, M. L., Kemler Nelson, D. G., Keil, F. C., & Gutierrez, F. (2006). What do children want to know about animals and artifacts? Domain-specific requests for information. *Psychological Science*, **17**, 455–9.

Guthrie, S. E. (1993). *Faces in the clouds. A new theory of religion*. New York: Oxford University Press.

Haeckel, E. (1886). *The evolution of man. A popular exposition of the principal points of human ontogeny and phylogeny*. New York: D. Appleton and Company.

Harmand, S., Lewis, J. E., Feibel, C. S., et al. (2015). 3.3-million-year-old stone tools from Lomekwi 3, West Turkana, Kenya. *Nature*, **521**, 310–5.

Harris, M. (2013). *The nature of creation. Examining the Bible and science.* Durham: Acumen.

Harrison, P. (1995). Newtonian science, miracles, and the laws of nature. *Journal of the History of Ideas*, **56**, 531–53.

Harrison, P. (2007). *The fall of man and the foundations of science.* Cambridge: Cambridge University Press.

Harvati, K, Röding, C., Bosman, A. M. et al. (2019). Apidima Cave fossils provide earliest evidence of *Homo sapiens* in Eurasia. *Nature*, **571**, 500–4.

Hick, J. (1966). *Evil and the god of love.* London: Macmillan.

Hill, K. R., Wood, B. M., Baggio, J., et al. (2014). Hunter-gatherer inter-band interaction rates: Implications for cumulative culture. *PLoS One*, **9**, e102806.

Hoffmann, D. L., Angelucci, D. E., Villaverde, V., et al. (2018). Symbolic use of marine shells and mineral pigments by Iberian Neandertals 115,000 years ago. *Science Advances*, **4**, eaar5255.

Hoffmann, D. L., Standish, C., García-Diez, M., et al. (2018). U-Th dating of carbonate crusts reveals Neandertal origin of Iberian cave art. *Science*, **359**, 912–5.

Holroyd, J. (2012). Responsibility for implicit bias. *Journal of Social Philosophy*, **43**, 274–306.

Hovers, E., & Belfer-Cohen, A. (2013). Insights into early mortuary practices of *Homo*. In S. Tarlow & L. N. Stutz (eds.), *The Oxford handbook of the archaeology of death and burial* (pp. 631–42). Oxford: Oxford University Press.

Hublin, J.-J. (2000). Modern–nonmodern hominid interactions: A Mediterranean perspective. In O. Bar-Yosef & D. Pilbeam (eds.), *The geography of the neandertals and modern humans in Europe and the greater Mediterranean* (pp. 157–82). Cambridge, MA: Harvard Peabody Museum.

Hume, D. (1748). *Philosophical essays concerning human understanding.* London: A. Millar.

Hume, D. (1757 [2007]). The natural history of religion. In T. L. Beauchamp (ed.), *A dissertation on the passions. The natural history of religion. A critical edition* (pp. 30–87). Oxford: Clarendon Press.

Huxley, T. H. (1863). *Evidences as to man's place in nature.* London: Williams and Norgate.

Irenaeus. (second century [1884]). *Against heresies* (A. Roberts & W. H. Rambaut, trans.). Edinburgh: T & T Clark.

Irenaeus. (second century [1997]). *On the apostolic preaching* (J. Behr, trans.). Crestwood, NY: St Vladimir's Seminary Press.

Ivanhoe, P. J. (2017). *Oneness. East Asian conceptions of virtue, happiness, and how we are all connected.* New York: Oxford University Press.

Jaeger, L. (2008). The idea of law in science and religion. *Science & Christian Belief*, **20**, 133–46.

Jaeger, L. (2017). Models of the Fall including a historical Adam as ancestor of all humans: Scientific and theological constraints. *Science & Christian Belief*, **29**, 20–36.

Järnefelt, E., Canfield, C. F., & Kelemen, D. (2015). The divided mind of a disbeliever: Intuitive beliefs about nature as purposefully created among different groups of non-religious adults. *Cognition*, **140**, 72–88.

Järnefelt, E., Zhu, L., Canfield, C. F., et al. (2019). Reasoning about nature's agency and design in the cultural context of China. *Religion, Brain & Behavior*, **9**, 156–78.

Johnson, E. A. (1996). Does God play dice? Divine providence and chance. *Theological Studies*, **57**, 3–18.

Kant, I. (1790 [2000]). *Critique of the power of judgment* (P. Guyer, ed., P. Guyer & E. Matthews, trans.). Cambridge: Cambridge University Press.

Kaplan, M. M. (1934 [2010]). *Judaism as a civilization. Towards a reconstruction of American-Jewish life*. Philadelphia: Jewish Publication Society.

Kauffman, S. (2008). *Reinventing the sacred. A new view of science, reason, and religion*. New York: Basic Books.

Kelemen, D. (1999). Why are rocks pointy? Children's preference for teleological explanations of the natural world. *Developmental Psychology*, **35**, 1440–52.

Kelemen, D. (2004). Are children "intuitive theists"? Reasoning about purpose and design in nature. *Psychological Science*, **15**, 295–301.

Kelemen, D., & Rosset, E. (2009). The human function compunction: Teleological explanation in adults. *Cognition*, **111**, 138–43.

Kelemen, D., Rottman, J., & Seston, R. (2013). Professional physical scientists display tenacious teleological tendencies: Purpose-based reasoning as a cognitive default. *Journal of Experimental Psychology: General*, **142**, 1074–83.

Kind, C.-J., Ebinger-Rist, N., Wolf, S., et al. (2014). The smile of the Lion Man. Recent excavations in Stadel Cave (Baden-Württemberg, southwestern Germany) and the restoration of the famous Upper Palaeolithic figurine. *Quartär*, **61**, 129–45.

Kuhn, S. L. (2014). Signaling theory and technologies of communication in the Paleolithic. *Biological Theory*, **9**, 42–50.

Lamarck, J.-B. (1809). *Philosophie zoologique, ou exposition des considérations relatives à l'histoire naturelle des animaux*. Paris: Duminil-Lesueur.

Lamoureux, D. O. (2008). *Evolutionary creation. A Christian approach to evolution*. Cambridge: Lutterworth Press.

Lamoureux, D. O. (2015). Beyond original sin: Is a theological paradigm shift inevitable? *Perspectives on Science and Christian Faith*, **67**, 35–49.

Landau, M. (1991). *Narratives of human evolution*. New Haven, CT: Yale University Press.

Legare, C. H., Evans, E. M., Rosengren, K. S., & Harris, P. L. (2012). The coexistence of natural and supernatural explanations across cultures and development. *Child Development*, **83**, 779–93.

Leibowitz, Y. (1992). *Judaism, human values, and the Jewish state* (E. Goldman, ed.). Cambridge, MA: Harvard University Press.

Leonard, W. R., & Robertson, M. L. (1997). Rethinking the energetics of bipedality. *Current Anthropology*, **38**, 304–9.

Levy, N. (2015). Neither fish nor fowl: Implicit attitudes as patchy endorsements. *Noûs*, **49**, 800–23.

Lewis-Williams, D. (2002). *The mind in the cave: Consciousness and the origins of art*. London: Thames & Hudson.

Lombrozo, T., Kelemen, D., & Zaitchik, D. (2007). Inferring design: Evidence of a preference for teleological explanations in patients with Alzheimer's disease. *Psychological Science*, **18**, 999–1006.

Lovelock, J. E., & Margulis, L. (1974). Atmospheric homeostasis by and for the biosphere: The Gaia hypothesis. *Tellus*, **26**, 2–10.

Maimonides, M. (twelfth century [1963]). *The guide of the perplexed* (S. Pines, trans.). Chicago: University of Chicago Press.

Martin, R. D., MacLarnon, A. M., Phillips, J. L., et al. (2006). Comment on "The brain of LB1, *Homo floresiensis*". *Science*, **312**, 999.

Mayr, E. (1992). The idea of teleology. *Journal of the History of Ideas*, **53**, 117–35.

McCall, G. S., & Shields, N. (2008). Examining the evidence from small-scale societies and early prehistory and implications for modern theories of aggression and violence. *Aggression and Violent Behavior*, **13**, 1–9.

McCauley, R. N. (2011). *Why religion is natural and science is not*. Oxford: Oxford University Press.

McDermott, R. (1970). The religion game: Some family resemblances. *Journal of the American Academy of Religion*, **38**, 390–400.

McGhee, G. (2011). *Convergent evolution. Limited forms most beautiful*. Cambridge, MA: Massachussetts Institute of Technology Press.

McGrath, A. E. (2011). *Darwinism and the divine. Evolutionary thought and natural theology*. Malden, MA: Wiley-Blackwell.

Miller, K. R. (1999 [2007]). *Finding Darwin's God: A scientist's search for common ground between God and evolution*. New York: Harper.

Milton, K. (1999). A hypothesis to explain the role of meat-eating in human evolution. *Evolutionary Anthropology*, **8**, 11–21.

Monod, J. (1970). *Le hasard et la nécessité. Essai sur la philosophie naturelle de la biologie moderne*. Paris: Éditions du Seuil.

Moon, A. (2017). Debunking morality: Lessons from the EAAN literature. *Pacific Philosophical Quarterly*, **98**, 208–26.

Murphy, N. (1995). Divine action in the natural order: Buridan's ass and Schrödinger's cat. In R. Russell, N. Murphy, & A. Peacocke (Eds.), *Chaos and complexity: Scientific perspectives on divine action* (pp. 325–58). Notre Dame: Vatican Observatory and Center for Theology and the Natural Sciences.

Murray, M. J. (2008). *Nature red in tooth and claw. Theism and the problem of animal suffering.* Oxford: Oxford University Press.

Neander, K. (1991). Functions as selected effects: The conceptual analyst's defense. *Philosophy of Science*, **58**, 168–84.

Nelson, S. (1990). Diversity of Upper Paleolithic 'Venus' figurines and archaeological mythology. In S. M. Nelson & A. B. Kehoe (eds.), *Powers of observation: Alternate views in archaeology* (pp. 11–22). Washington, DC: American Anthropological Association.

Newsom, C. A. (2003). *The book of Job: A contest of moral imaginations.* Oxford: Oxford University Press.

Norenzayan, A. (2013). *Big gods. How religion transformed cooperation and conflict.* Princeton, NJ: Princeton University Press.

Norenzayan, A., & Gervais, W. M. (2013). The origins of religious disbelief. *Trends in Cognitive Sciences*, **17**, 20–5.

O'Brien, M. J., & Bentley, R. A. (2015). The role of food storage in human niche construction: An example from Neolithic Europe. *Environmental Archaeology*, **20**, 364–78.

Okasha, S. (2018). *Agents and goals in evolution.* Oxford: Oxford University Press.

Paluck, E. L., Shepherd, H., & Aronow, P. M. (2016). Changing climates of conflict: A social network experiment in 56 schools. *Proceedings of the National Academy of Sciences USA*, **113**, 566–71.

Pedersen, D. (2017). *The eternal covenant: Schleiermacher on God and natural science.* New York and Berlin: De Gruyter.

Peoples, H. C., Duda, P., & Marlowe, F. W. (2016). Hunter-gatherers and the origins of religion. *Human Nature*, **27**, 261–82.

Pierce, J. (2013). The dying animal. *Journal of Bioethical Inquiry*, **10**, 469–78.

Plantinga, A. (2000). *Warranted Christian belief.* New York: Oxford University Press.

Plantinga, A. (2011). *Where the conflict really lies. Science, religion, and naturalism.* Oxford: Oxford University Press.

Poling, D. A., & Evans, E. M. (2004). Are dinosaurs the rule or the exception? Developing concepts of death and extinction. *Cognitive Development*, **19**, 363–83.

Potter, B. (1909). *The tale of the flopsy bunnies*. London: Frederick Warne & Co.

Pruvost, M., Bellone, R., Benecke, N., et al. (2011). Genotypes of predomestic horses match phenotypes painted in Paleolithic works of cave art. *Proceedings of the National Academy of Sciences USA*, **108**, 18626–30.

Purzycki, B. G., Apicella, C., Atkinson, Q. D., et al. (2016). Moralistic gods, supernatural punishment and the expansion of human sociality. *Nature*, **530**, 327–30.

Raup, D. M. (1991). *Extinction. Bad genes or bad luck?* New York: W. W. Norton.

Rauschenbusch, W. (1917). *A theology for the social gospel*. New York: MacMillan.

Reddish, P., Fischer, R., & Bulbulia, J. (2013). Let's dance together: Synchrony, shared intentionality and cooperation. *PloS One*, **8**, e71182.

Reeves, J. (2015). The secularization of chance: Toward understanding the impact of the probability revolution on Christian belief in divine providence. *Zygon: Journal of Religion and Science*, **50**, 604–20.

Reich, D., Green, R. E., Kircher, M., et al. (2010). Genetic history of an archaic hominin group from Denisova Cave in Siberia. *Nature*, **468**, 1053–60.

Reiss, J. O. (2009). *Not by design: Retiring Darwin's watchmaker*. Berkeley: University of California Press.

Richards, R. J. (2000). Kant and Blumenbach on the *Bildungstrieb*: A historical misunderstanding. *Studies in History and Philosophy of Biological and Biomedical Sciences*, **31**, 11–32.

Richerson, P. J., & Boyd, R. (2005). *Not by genes alone. How culture transformed human evolution*. Chicago: University of Chicago Press.

Richerson, P., & Henrich, J. (2012). Tribal social instincts and the cultural evolution of institutions to solve collective action problems. *Cliodynamics: Journal of Theoretical and Mathematical History*, **3**, 38–80.

Richter, D., Grün, R., Joannes-Boyau, R. et al. (2017). The age of the hominin fossils from Jebel Irhoud, Morocco, and the origins of the Middle Stone Age. *Nature*, **546**, 293–6.

Ritchie, S. L. (2017). Dancing around the causal joint: Challenging the theological turn in divine action theories. *Zygon: Journal of Religion and Science*, **52**, 361–79.

Roes, F. L., & Raymond, M. (2003). Belief in moralizing gods. *Evolution and Human Behavior*, **24**, 126–35.

Rolston, H. (2018). Redeeming a cruciform nature. *Zygon: Journal of Religion and Science*, **53**, 739–51.

Ruse, M. (2003). *Darwin and design. Does evolution have a purpose?* Cambridge, MA: Harvard University Press.

Ruse, M. (2016). Evolutionary biology and the question of teleology. *Studies in History and Philosophy of Biological and Biomedical Sciences*, **58**, 100–6.

Russell, R. J. (2013). Recent theological interpretations of evolution. *Theology and Science*, **11**, 169–84.

Sala, N., Arsuaga, J. L., Pantoja-Pérez, A., et al. (2015). Lethal interpersonal violence in the Middle Pleistocene. *PLoS ONE*, **10**, e0126589.

Sala, N., & Conard, N. (2016). Taphonomic analysis of the hominin remains from Swabian Jura and their implications for the mortuary practices during the Upper Paleolithic. *Quaternary Science Reviews*, **150**, 278–300.

Saladie, P., Huguet, R., Rodriguez-Hidalgo, A., et al. (2012). Intergroup cannibalism in the European Early Pleistocene: The range expansion and imbalance of power hypotheses. *Journal of Human Evolution*, **63**, 682–95.

Salali, G. D., Chaudhary, N., Thompson, J., et al. (2016). Knowledge-sharing networks in hunter-gatherers and the evolution of cumulative culture. *Current Biology*, **26**, 2516–21.

Salque, M., Bogucki, P. I., Pyzel, J., et al. (2013). Earliest evidence for cheese making in the sixth millennium BC in northern Europe. *Nature*, **493**, 522–5.

Schleiermacher, F. (1830 [2016]). *Christian Faith* (N. Tice, C. L. Kelsey, & E. Lawler, trans.). Louisville, KY: Westminster John Knox Press.

Schmidt, M. F., Butler, L. P., Heinz, J., & Tomasello, M. (2016). Young children see a single action and infer a social norm: Promiscuous normativity in 3-year-olds. *Psychological Science*, **27**, 1360–70.

Schneider, J. R. (in press). *Animal suffering and the Darwinian problem of evil*. Cambridge: Cambridge University Press.

Sedgwick, A. (1845 [1890]). Letter to Charles Lyell, April 9, 1845. In J. Willis Clark & T. McKenny Hughes (Eds.), *The life and letters of the Reverend Adam Sedgwick* (pp. 183–5). Cambridge: Cambridge University Press.

Sedley, D. (2007). *Creationism and its critics in antiquity*. Berkeley & Los Angeles: University of California Press.

Slifkin, N. (2008). *The challenge of creation: Judaism's encounter with science, cosmology, and evolution*. Jerusalem: Yashar Books.

Smith, J. (2017). What stands on the Fall? A philosophical exploration. In W. Cavanaugh & J. Smith (eds.), *Evolution and the Fall* (pp. 48–64). Grand Rapids, MI: Eerdmans.

Sosis, R., & Bressler, E. R. (2003). Cooperation and commune longevity: A test of the costly signaling theory of religion. *Cross-Cultural Research*, **37**, 211–39.

Sosis, R., & Kiper, J. (2014). Religion is more than belief: What evolutionary theories of religion tell us about religious commitments. In M. Bergmann &

P. Kain (eds.), *Challenges to moral and religious belief. Disagreement and evolution* (pp. 256–76). Oxford: Oxford University Press.

Sosis, R., & Ruffle, B. J. (2007). Religious ritual and cooperation: Testing for a relationship on Israeli religious and secular kibbutzim. *Current Anthropology*, **44**, 713–22.

Southgate, C. (2008). *The groaning of creation. God, evolution, and the problem of evil.* Louisville, KY: Westminster John Knox Press.

Štrkalj, G. (2003). Robert Broom's theory of evolution. *Transactions of the Royal Society of South Africa*, **58**, 35–9.

Swammerdam, J. (1669). *Historia insectorum generalis, ofte algemeene verhandeling van de bloedeloose dierkens.* Utrecht: Meinardus van Dreunen.

Tattersall, I. (1998). *Becoming human: Evolution and human uniqueness.* Oxford: Oxford University Press.

Tenesa, A., Navarro, P., Hayes, B. J., et al. (2007). Recent human effective population size estimated from linkage disequilibrium. *Genome Research*, **17**, 520–6.

Tennant, F. (1902). *The origin and propagation of sin.* Cambridge: Cambridge University Press.

Tennant, F. R. (1903). *The sources of the doctrines of the Fall and original sin.* Cambridge: Cambridge University Press.

Tennant, F. R. (1912). *The concept of sin.* Cambridge: Cambridge University Press.

van den Bergh, G. D., Kaifu, Y., Kurniawan, I., et al. (2016). *Homo floresiensis*-like fossils from the early Middle Pleistocene of Flores. *Nature*, **534**, 245–9.

van den Toren, B. (2016). Human evolution and a cultural understanding of original sin. *Perspectives on Science and Christian Faith*, **68**, 12–22.

Van Gelder, L. (2015). Counting the children: The role of children in the production of finger flutings in four Upper Palaeolithic caves. *Oxford Journal of Archaeology*, **34**, 119–38.

van Huyssteen, W. J. (2006). *Alone in the world? Human uniqueness in science and theology.* Göttingen: Vandenhoeck & Ruprecht.

Ward, K. (1996). *God, chance and necessity.* Oxford: Oneworld.

Watson-Jones, R. E., Busch, J. T., Harris, P. L., & Legare, C. H. (2017). Does the body survive death? Cultural variation in beliefs about life everlasting. *Cognitive Science*, **41**, 455–76.

Watts, I., Chazan, M., Wilkins, J., et al. (2016). Early evidence for brilliant ritualized display: Specularite use in the Northern Cape (South Africa) between 500 and 300 ka. *Current Anthropology*, **57**, 287–310.

Watts, J., Greenhill, S. J., Atkinson, Q. D., et al. (2015). Broad supernatural punishment but not moralizing high gods precede the evolution of political complexity in Austronesia. *Proceedings of the Royal Society of London B: Biological Sciences*, **282**, 20142556.

White, C. (2018). What does the cognitive science of religion explain? In H. van Eyghen, R. Peels, & G. van den Brink (Eds.), *New developments in the cognitive science of religion* (pp. 35–49). Dordrecht: Springer.

White, R. (1993). A social and technological view of Aurignacian and Castelperronian personal ornaments in SW Europe. In V. Cabrera Valdés (ed.), *El origen del hombre moderno en el suroeste de Europa* (pp. 327–57). Madrid: Ministerio de Educación y Ciencia.

White, T. D. (1995). African omnivores: Global climatic change and Plio-Pleistocene hominids and suids. In E. S. Vrba, G. H. Denton, T. C. Partridge, & L. H. Burckle (eds.), *Paleoclimate and evolution, with emphasis on human origins* (pp. 369–84). New Haven, CT: Yale University Press.

Wiessner, P. (2002). Hunting, healing, and *hxaro* exchange: A long-term perspective on !Kung (Ju/'hoansi) large-game hunting. *Evolution and Human Behavior*, **23**, 407–36.

Williams, P. A. (2001). *Doing without Adam and Eve: Sociobiology and original sin*. Minneapolis: Fortress Press.

Wolff, C. F. (1759). *Theoria generationis*. Halle: Litteris Hendelianis.

Xenophanes. (sixth to fifth century BCE [1898]). *The first philosophers of Greece* (A. Fairbanks, ed.). London: Kegan Paul, Trench, Trübner and Co.

Zhivotovsky, L. A., Rosenberg, N. A., & Feldman, M. W. (2003). Features of evolution and expansion of modern humans, inferred from genomewide microsatellite markers. *American Journal of Human Genetics*, **72**, 1171–86.

Zilhão, J. (2012). Personal ornaments and symbolism among the Neanderthals. *Developments in Quaternary Sciences*, **16**, 35–49.

Acknowledgments

We thank Daniel Pedersen, Samuel Lebens, Alexus McLeod, an anonymous reviewer for Cambridge University Press, Michael Ruse, and Grant Ramsey, as well as audiences at Fuller Theological Seminary and the Catholic University of Lyon for their helpful comments. This research was supported by the John Templeton Foundation, grant ID 60802.

Cambridge Elements ≡

Philosophy of Biology

Grant Ramsey

KU Leuven

Grant Ramsey is a BOFZAP research professor at the Institute of Philosophy, KU Leuven, Belgium. His work centers on philosophical problems at the foundation of evolutionary biology. He has been awarded the Popper Prize twice for his work in this area. He also publishes in the philosophy of animal behavior, human nature, and the moral emotions. He runs the Ramsey Lab (theramseylab.org), a highly collaborative research group focused on issues in the philosophy of the life sciences.

Michael Ruse

Florida State University

Michael Ruse is the Lucyle T. Werkmeister Professor of Philosophy and the Director of the Program in the History and Philosophy of Science at Florida State University. He is Professor Emeritus at the University of Guelph, in Ontario, Canada. He is a former Guggenheim fellow and Gifford lecturer. He is the author or editor of over sixty books, most recently *Darwinism as Religion: What Literature Tells Us about Evolution*; *On Purpose*; *The Problem of War: Darwinism, Christianity, and their Battle to Understand Human Conflict*; and *A Meaning to Life*.

About the Series

This Cambridge Elements series provides concise and structured introductions to all of the central topics in the philosophy of biology. Contributors to the series are cutting-edge researchers who offer balanced, comprehensive coverage of multiple perspectives, while also developing new ideas and arguments from a unique viewpoint.

Cambridge Elements $\equiv$

Philosophy of Biology

9 781108 716048